Conversations on Electric
and Magnetic Fields in the Cosmos

PRINCETON SERIES IN ASTROPHYSICS
EDITED BY DAVID N. SPERGEL

1. Theory of Rotating Stars, *by Jean-Louis Tassoul*
2. Theory of Stellar Pulsation, *by John P. Cox*
3. Galactic Dynamics, *by James Binney and Scott Tremaine*
4. Dynamical Evolution of Globular Clusters, *by Lyman Spitzer, Jr.*
5. Supernovae and Nucleosynthesis: An Investigation of the History of Matter, from the Big Bang to the Present, *by David Arnett*
6. Unsolved Problems in Astrophysics, *edited by John N. Bahcall and Jeremiah P. Ostriker*
7. Galactic Astronomy, *by James Binney and Michael Merrifield*
8. Active Galactic Nuclei: From the Central Black Hole to the Galactic Environment, *by Julian H. Krolik*
9. Plasma Physics for Astrophysics, *by Russell M. Kulsrud*
10. Electromagnetic Processes, *by Robert J. Gould*
11. Conversations on Electric and Magnetic Fields in the Cosmos, *by Eugene N. Parker*

Conversations on Electric and Magnetic Fields in the Cosmos

Eugene N. Parker

PRINCETON UNIVERSITY PRESS
PRINCETON AND OXFORD

Published by Princeton University Press, 41 William Street, Princeton, New Jersey 08540
In the United Kingdom: Princeton University Press, 3 Market Place, Woodstock, Oxfordshire OX20 1SY

Library of Congress Cataloging-in-Publication Data

Parker, E. N. (Eugene Newman), 1927-
Conversations on electric and magnetic fields in the cosmos / Eugene N. Parker.
p. cm. – (Princeton series in astrophysics)
Includes bibliographical references and index.
ISBN-13: 978-0-691-12840-5 (cloth : alk. paper)
ISBN-10: 0-691-12840-5 (cloth : alk. paper)
ISBN-13: 978-0-691-12841-2 (pbk. : alk. paper)
ISBN-10: 0-691-12841-3(pbk. : alk. paper)

1. Plasma electrodynamics. 2. Electromagnetic fields. 3. Maxwell equations. 4. Electromagnetic theory. I. Title.
QC718.5.E4P37 2007
523.01'87—dc22

2006024259

British Library Cataloging-in-Publication Data is available

This book has been composed in Sabon

Printed on acid-free paper. ∞

pup.princeton.edu

Printed in the United States of America

10 9 8 7 6 5 4 3 2 1

To Niesje

Contents

Illustrations

Acknowledgments

I am grateful to several colleagues whose comments and suggestions have been invaluable in clarifying the narrative at crucial points and in rounding out some of the topics that were treated too briefly in the first draft. I thank N. O. Weiss of the Department of Applied Mathematics and Theoretical Physics, Cambridge University for stimulating discussion. I appreciate the extensive discussions with Vytenis Vasyliunas of the Max Planck Institut für Sonnensystemforschung, Lindau. Some of his own work forms an important complement to the ideas expressed here, and is referenced accordingly throughout these *Conversations*. The anonymous reviewer for Princeton University Press pointed out several places in the text that needed clarification. The author is too close to the narrative to maintain adequate perspective, and the fresh look of an "outsider" is essential to pick up the ambiguities in the seemingly "crystal clear" words of the initial writing. In particular, I thank B. C. Low and T. J. Bogdan of the High Altitude Observatory, NCAR, Boulder, Colorado for their critical readings of the manuscript and their comments thereon. They were invaluable in smoothing the flow of ideas and in picking up on the related concepts bypassed in the first writing.

Conversations on Electric and Magnetic Fields in the Cosmos

1 Introduction

1.1 General Remarks

The theoretical structure of electric and magnetic fields is presented in the standard textbooks, and one may ask why further conversation on the subject is useful or interesting. What is new that has not already been said many times before? The reply is that the emphasis in the usual formulation of electromagnetism is directed toward static electric and magnetic fields and then to electromagnetic radiation, whereas we are interested here in the electromagnetism of the cosmos—the large-scale magnetic fields that are transported bodily in the swirling ionized gases (plasmas) of planetary magnetospheres, stars, and galaxies, and, indeed, throughout intergalactic space. The plasma and the magnetic fields appear to be everywhere throughout the universe. The essential feature is that no significant electric field can arise in the frame of reference of the moving plasma. Hence, the large-scale dynamics of the magnetic field is tied to the hydrodynamics (HD) of the swirling plasma in the manner described by theoretical magnetohydrodynamics (MHD). So we shall have a fresh look at the theoretical foundations of both HD and MHD. The conventional derivations of the basic equations of HD and MHD are correct, of course, but the derivations ignore some fundamental questions, allowing a variety of misconceptions to flourish in the scientific community. We work out a minimal physical derivation, laying bare the simplicity of the necessary and sufficient conditions for the validity of HD and MHD to describe the large-scale bulk motion of plasmas and their magnetic fields. The essential condition for HD is that there be enough particles to give a statistically precise definition of the local plasma density; the essential condition for MHD is that there be enough free electrons and ions that the plasma cannot support any significant electric field in its own moving frame of reference. Both of these requirements are satisfied almost everywhere throughout the cosmos, with the result that HD and MHD accurately describe the large-scale bulk dynamics of the plasmas and fields. The magnetic field is transported bodily with the bulk motion of the plasma, and the dynamics is basically the mechanical interaction between the stresses in the magnetic field $\mathbf{B}$ and the pressure p_{ij} and bulk momentum density $NM\mathbf{v}$ of the plasma velocity $\mathbf{v}$. The associated electric current $\mathbf{j}$ and the electric field $\mathbf{E}$ in the laboratory frame of reference play no direct role in the dynamics. They are created and driven by the varying $\mathbf{B}$ and $\mathbf{v}$. If needed for some purpose, they are readily computed once the dynamics has provided $\mathbf{B}$ and $\mathbf{v}$.

It is here that a fundamental misunderstanding has become widely accepted, mistaking the electric current j and the electric current **E** (the E, j paradigm) (Parker 1996a) to be the fundamental physical entities. Steady conditions often can be treated using the E, j paradigm, but the dynamics of time-dependent systems becomes difficult, if not impossible, because of the inability to express Newton's equation in terms of E and j in a tractable form. That is to say, E and j are proxies for **B** and **v**, but too remote from **B** and **v** to handle the momentum equation. So it is not possible to construct a workable set of dynamical field equations in terms of j and E from the equations of Newton and Maxwell. The generalized Ohm's law is often employed, but Ohm's law does not control the large-scale dynamics. The tail does not wag the dog. This inadequacy has led to fantasy to complement the limited equations available in the E, j paradigm, attributing the leading dynamical role to an electric field **E** with unphysical properties. Magnetospheric physics has suffered severely from this misdirection, and we will come back to the specific aspects of the misunderstanding at appropriate places in these conversations.

The essential point is that we live in a magnetohydrodynamic universe in which the magnetic field **B** is responsible for the remarkable behavior of the gas velocity **v**, and vice versa. Then we must recognize that the large-scale magnetic stresses in the interlaced field line topologies created by the plasma motions have the peculiar property of causing the field gradients to increase without bound. The resulting thin layers of intense field shear and high current density "eat up" the magnetic fields at prodigious rates. The effect is commonly called *rapid reconnection* of the magnetic field because the field lines are cut and rejoined across the intense shear layer, and it is a universal consequence of the large-scale field line topology. Rapid reconnection is evidently responsible for such phenomena as the solar flare, the million degree temperature of the solar X-ray corona, and the terrestrial aurora. So the MHD universe is far more active and interesting than a purely HD universe, with the magnetic activity of the Sun an outstanding example. R. W. Leighton remarked many years ago that if it were not for magnetic fields, the Sun would be as uninteresting as most astronomers seem to think it is. The activity of the Sun is the model, then, for the unresolved activity of other stars.

The conversation is intended to complement, rather than replace, the familiar textbook development of electromagnetic theory and of HD and MHD. It is assumed that the reader is already familiar with the conventional development of electromagnetic theory, and it is to be hoped that the reader has the patience to follow the conversation when it briefly reiterates some of that familiar boilerplate, because the basics are necessarily the same, even as we provide a different emphasis.

There will be some new twists to the development along with the boilerplate. For instance, we show that the Biot-Savart integral form of Ampere's law implies Maxwell's equation. This will-o'-the-wisp is rediscovered every decade or so, but never seems to get into the standard textbooks. It has amusing implications for the early controversy over Maxwell's equation. Then we point out the singular properties of the Maxwell stress tensor in arbitrary equilibrium field topologies.

We show that the familiar equations of hydrodynamics are required by the principles of conservation of particles, momentum, and energy in the large-scale bulk flow of the plasma. These are valid principles regardless of the presence or absence of interparticle collisions and magnetic fields. As already noted, HD is valid so long as there are enough particles to provide a statistically well-defined fluid density, contrary to what one sometimes reads in the literature about the relatively collisionless plasma. We show, too, that the familiar equations of magnetohydrodynamics are inescapable unless there are so few free electrons and ions that the gas is an effective electrical insulator. The air that we breathe is an example, and only upon reaching the ionosphere does MHD become effective.

In particular, the conversation emphasizes the principle—Occam's razor—that the theoretical concepts should contain no unnecessary embellishments. So we prune away concepts and notation that are not vital to the experimental physics, and we note in particular that physical reality is made up of the manner in which things are experimentally perceived to be. This seemingly trivial point is commonly violated by the vocabulary of magnetic induction, and it leads us into conflict with a variety of customs and popular opinions.

The reader will soon see that the conversation enters into numerous digressions, examining and commenting on the scenery as we pass along the minimum theory road. The writing of minimal theory is not obligated to provide only the minimum conversation.

1.2 Electromagnetic Field Equations

Our cosmos exhibits some remarkable electromagnetic symmetries and some remarkable electromagnetic asymmetries, and it is interesting to have a look at both. We begin by noting the well-known fact that, on the one hand, the electric and magnetic fields, **E** and **B**, respectively, are equal partners in their interactions, described by Maxwell's equations

$$\frac{\partial \mathbf{B}}{\partial t} = -c\nabla \times \mathbf{E} \tag{1.1}$$

$$\frac{\partial \mathbf{E}}{\partial t} = +c\nabla \times \mathbf{B} \tag{1.2}$$

in a vacuum. These two equations, symmetric in **E** and **B**, state simply that any change in either field with the passage of time is accompanied by a proportionate curl of the other, and vice versa. That is the nature of the electromagnetic wave, so it should be no surprise that the proportionality constant, having dimensions of velocity, turns out to be the speed of light c.

Now the **E**, **B** symmetry of the field equations is in contrast with the fact that the universe is itself unsymmetric with respect to electric and magnetic charges. The matter throughout the universe is found to consist only of electrically charged particles, i.e., electrons and nucleons, with no indication of magnetic charges. Obviously, the cosmos was not created by gravity and electromagnetic forces alone.

The reader can see that the conversation employs cgs units, or their equivalent, rather than SI or mks units. The motive is to exhibit the basic dynamical symmetry between **E** and **B**, so thoroughly obscured by SI units in which **E** and **B** are assigned different dimensions! As discussed in section 6.4, the SI treatment, insisting upon the coulomb as the unit of charge, introduces superfluous concepts, contrary to the principle of minimum theoretical complexity.

Now the fact that most of the gases in the universe are at least partially ionized means an abundance of free electrons and ions. Hence, the electric current density **j** is created by a very weak electric field, quickly reducing any large-scale electric field **E**′ in the frame of reference of the moving plasma to negligible values.

There can be no magnetic current **J** because there are no magnetic charges—magnetic *monopoles*—so far as anyone can tell. However, it is not without interest to look briefly into the physical consequences of an abundance of monopoles—a monopole plasma. Maxwell's equations would be written

$$4\pi\mathbf{J} + \frac{\partial \mathbf{B}}{\partial t} = -c\nabla \times \mathbf{E} \qquad \nabla\cdot\mathbf{B} = 4\pi\Delta \tag{1.3}$$

$$4\pi\mathbf{j} + \frac{\partial \mathbf{E}}{\partial t} = +c\nabla \times \mathbf{B} \qquad \nabla\cdot\mathbf{E} = 4\pi\delta \tag{1.4}$$

where Δ is the magnetic charge density and δ is the electric charge density. Both the electric current density **j** and the magnetic current density **J** have been introduced into the left-hand side of the vacuum equations. Presumably, the cosmos would have no total magnetic charge, just as we commonly suppose it has no total electric charge, although we will come back to this point later to consider the speculations of Lyttleton and Bondi (1959, 1960; Hoyle 1960).

Assuming that the individual magnetic monopoles are as mobile as the free electrons and ions, the magnetic monopole plasma would reduce the magnetic field in its own moving frame of reference to negligible values, just as the ion–electron plasma eliminates an electric field in its own frame of reference. The nonrelativistic Lorentz transformations between the electric and magnetic fields **E** and **B** in the laboratory and **E**′ and **B**′ in the reference frame moving with velocity **v** relative to the laboratory are

$$\mathbf{E}' = \mathbf{E} + \frac{\mathbf{v} \times \mathbf{B}}{c} \qquad \mathbf{B}' = \mathbf{B} - \frac{\mathbf{v} \times \mathbf{B}}{c}$$

Suppose, then, that the ion–electron plasma has a velocity **v** relative to the laboratory. The free electrons reduce **E**′ in that plasma to zero, from which it follows that

$$\mathbf{E} = \frac{-\mathbf{v} \times \mathbf{B}}{c}$$

in the laboratory. Similarly, suppose that the magnetic monopole plasma has a velocity **V** relative to the laboratory, and the free monopoles reduce **B**′ in that plasma to zero. Then

$$\mathbf{B} = \frac{+\mathbf{V} \times \mathbf{E}}{c}$$

in the laboratory. When **B** is eliminated between these two equations, it follows that

$$\begin{aligned}\mathbf{E} &= \frac{-\mathbf{v} \times (\mathbf{V} \times \mathbf{E})}{c^2}\\ &= -\frac{\mathbf{V}(\mathbf{v}\cdot\mathbf{E})/c^2}{1 - \mathbf{V}\cdot\mathbf{v}/c^2}\end{aligned}$$

Hence, **E** is parallel to **V**, and it follows that **B** = 0. Eliminating **E** between the two equations yields

$$\mathbf{B} = -\frac{\mathbf{v}(\mathbf{V}\cdot\mathbf{B})/c^2}{1 - \mathbf{V}\cdot\mathbf{v}/c^2}$$

So **B** is parallel to **v**, from which it follows that **E** = 0. So that universe would be hydrodynamic (HD) rather than magnetohydrodynamic

(MHD), and the Sun would not show anything comparable to its present MHD activity.

There would, of course, be some interesting things to do with magnetic monopoles. For instance, a toothpick with an electric charge attached to one end and a magnetic charge to the other forms a system with net electromagnetic angular momentum, given the electromagnetic momentum density $\mathbf{E} \times \mathbf{B}/4\pi c$. The toothpick would represent a gyroscope with no moving parts. However, that would be small compensation for the absence of relativistic jets, double radio sources, synchrotron emission, cosmic rays, etc., that are to be found in our own cosmos. There would be no sunspots, no prominences, no flares, no corona, no coronal mass ejections, no solar wind, no geomagnetic field, and no magnetic compass, to name but a few of the things missing from that universe.

In fact, given the central role of magnetic fields in determining the nature of the accretion disks involved in the formation of stars, one may ask to what extent there would be stars and planets? And the possibility of life? It is fashionable these days to conjecture on the existence of parallel universes. So we remark that there might be another universe "somewhere" with both mobile electric charges and mobile magnetic charges with no one in that cosmos to contemplate it.

Another question that springs to mind is whether there might exist a universe in which only magnetic charges exist, so that it would be a replica of our own cosmos except that the atoms would consist of light magnetically charged particles clustered around oppositely charged massive magnetic particles. Communicating by radio between the two universes, would it be possible to determine the difference?

Whether such a universe exists depends on the properties of the fundamental particles in other universes. Contemporary particle theory in our own universe suggests that, if they exist, magnetic monopoles have a mass μ at least as large as 10^{16} GeV, or about 10^{-8} g. There are living organisms with substantially less mass than that. With such a mass, the monopole, with a magnetic charge $g = 137e/2$, is not what we would call mobile. The acceleration in a large-scale magnetic field $\mathbf{B}$ is equal to $g\mathbf{B}/\mu$. In the large-scale magnetic field $B = 4 \times 10^{-6}$ G in a galaxy it would require 4×10^4 years to accelerate the monopole to 100 km/s. That opens up the possibility that a universe with such massive monopoles might be subject to magnetic monopole plasma oscillations, with a monopole plasma frequency $\Omega = (4\pi n g^2/\mu)^{1/2}$, where n is the monopole number density (Turner et al. 1982). This question comes up again in section 1.4, where we review the upper limits on n imposed in our universe by the existence of the magnetic field of the Galaxy.

On the other hand, if one imagines that the other universe forms the magnetic analog of our universe of electrically charged particles, then one would have magnetic monopoles with a magnetic charge e and the small mass m of the electron, and there would be oppositely charged monopoles with the mass M of a proton. Theory in that universe would suggest the possibility of electrically charged particles with masses of the order of 10^{16} GeV, etc.

Large-scale magnetic fields in that magnetic monopole universe would be quickly neutralized in the frame of reference of the swirling monopole plasma, in direct analogy to the obliteration of electric fields in the frame of reference of the plasma in our own universe. The monopole universe would be filled with large-scale electric fields tied to the hydrodynamics of the swirling monopole plasma, in the manner described by theoretical "electrohydrodynamics" (EHD)—the exact analog of MHD.

Returning to the realities of our own cosmos, it appears that we may neglect the occasional magnetic monopole, if there are any at all, and Maxwell's equations are written as

$$\frac{\partial \mathbf{B}}{\partial t} = -c\nabla \times \mathbf{E} \qquad \nabla \cdot \mathbf{B} = 0 \tag{1.5}$$

$$4\pi \mathbf{j} + \frac{\partial \mathbf{E}}{\partial t} = +c\nabla \times \mathbf{B} \qquad \nabla \cdot \mathbf{E} = 4\pi\delta \tag{1.6}$$

Note, then, that these electromagnetic equations make contact with Newtonian mechanics through the electric charge density δ and the mechanical motion of the charges associated with the current density $\mathbf{j}$, on which more will be said in chapter 6.

1.3 Electrical Neutrality

It is important to appreciate the large charge to mass ratio of the electron ($e/m = 5.3 \times 10^{17}$ cgs, compared to g/M = 2cgs for the theoretical magnetic monopole). Thus, for instance, one volt of potential difference accelerates an electron to 600 km/s. Note then that one mole of free electrons (6×10^{23} electrons) has a mass of about half a milligram and a charge of 3×10^{14} cgs (10^5 C), enough to supply a current of 1 A for a day. An electric charge of this amount would raise the Sun (radius 7×10^{10} cm) to about 1.3×10^6 V. If the half-milligram of electrons were then released, they would accelerate away from the Sun to relativistic velocities (the rest mass of an electron is 0.5×10^6 eV). Noting again that a potential difference of 1 V is sufficient to accelerate an electron to 600 km/s, it is evident that even a very weak electric field applied to a plasma would produce an immense current. That is to say, the electric field is limited to very small values by the highly charged and freely moving electrons.

As an illustrative example, consider the simple case of a plasma sufficiently dense that the scalar Ohm's law applies, so that

$$\mathbf{j} = \sigma \mathbf{E}' \tag{1.7}$$

where $\mathbf{E}'$ is the electric field in the frame of reference moving with the local plasma. For ionized hydrogen the electrical conductivity is $\sigma = 2 \times 10^7 T^{3/2}$/s (cf. Spitzer 1956). Starting with Ampere's law $4\pi\mathbf{j} = c\nabla \times \mathbf{B}$, it follows that $4\pi j \approx cB/l$, in order of magnitude, where l is the characteristic scale of variation of the magnetic field $\mathbf{B}$. Then with $\mathbf{E}' = \mathbf{j}/\sigma$, it follows that the field magnitudes are related by

$$\frac{E'}{B} = \frac{c}{4\pi\sigma l} \tag{1.8}$$

$$= \frac{10^{-4}}{l}\left(\frac{10^4}{T}\right)^{3/2} \tag{1.9}$$

in order of magnitude, with l in centimeters. Ionized hydrogen suggests $T \geq 10^4$ K. So with l as small as 1 km it follows that $E'/B \leq 10^{-9}$. On the large scales associated with stars and galaxies, E'/B is very small indeed. It is evident, then, that electric field stresses are insignificant compared to magnetic field stresses, the ratio of the stresses being $(E'/B)^2$. We will have more to say on this later. Only on the very small scales arising in shock fronts and in spontaneous tangential discontinuities and rapid magnetic reconnection can the electric field play a significant role in the dynamics.

Consider, then, the curious fact that one sign of charge—the one we call negative—is associated with the lightweight electron or lepton, while the opposite charge—called positive—is associated with the proton or baryon. Why such different particles? Perhaps the answer lies in the experimental fact that the positive and negative manifestations of charge assigned to the same type of particle represent a particle and its antiparticle. Such particles annihilate when they meet and convert to photons—gamma rays. So there have to be two different classes of stable particle if there is to be long-lived matter. Positronium just does not do it by itself. Our universe possesses the stable leptons and baryons, and here we are to take note of it.

That said, note that *positive* charge is defined as the charge left on a glass rod after rubbing the rod with silk, the silk being charged negatively, by definition, with the electrons rubbed off the rod. With the same rule applied in an anti-universe, where the atoms consist of positrons orbiting antiprotons, the anti-glass rod could just as well be given the positive charge appellation by the anti-people that live there. The question arises as to how a physicist and an anti-physicist in radio communication would determine whether their universes were both made up of matter, or whether one universe was the anti-universe of the other. Needless to say, if they could meet in person somehow, a handshake would quickly settle the matter. Such a cordial meeting would be the ultimate game of Russian roulette.

The issue could be settled via the radio if the two could (a) establish a common direction perpendicular to the direction of propagation of the radio signals employed for their communication and then (b) agree on a positive and negative sense along that direction. Whether this can be achieved depends on the nature of the radio access between the two universes, on which we can say no more. If they cannot achieve (b), then they cannot know the answer to their question. However, if the anti-physicist resides on an anti-matter planet in our own familiar cosmos—Alfvèn (1966) proposed that half the planets and stars in the Galaxy are matter and half are anti-matter—then the communication can decide the issue, at least in principle. The direction (b) could be decided by observing a common object, e.g., a distinctive distant galaxy, in a direction perpendicular to the line of propagation of their radio signal, and assigning a positive direction along that line, referred to as the positive (b) direction in the sequel.

Then, to proceed with the measurements, suppose that at time $t = 0$ our physicist displaces a charge $+q$ with a modest velocity $\mathbf{v}(t)$ ($<< c$). The vector potential of the radiation at the distant location $\mathbf{r}$ would be

$$\mathbf{A}(r, t) = \frac{q\mathbf{v}(t - r/c)}{cr}$$

The associated electric field is

$$\begin{aligned} \mathbf{E} &= \frac{q}{c^2 r}\left[\mathbf{a}\left(t - \frac{r}{c}\right) \times \mathbf{n}\right] \times \mathbf{n} \\ &= -\frac{q}{c^2 r}\mathbf{a}_{\perp}\left(t - \frac{r}{c}\right) \end{aligned}$$

where $\mathbf{a}(t)$ is the acceleration $d\mathbf{v}/dt$, $\mathbf{n}$ is the unit vector in the radial direction ($\mathbf{r} = \mathbf{n}r$), and $\mathbf{a}_{\perp}$ designates the component of $\mathbf{a}$ perpendicular to $\mathbf{r}$. The acceleration $\mathbf{a}$ is described by radio to the distant physicist in terms of the (b) direction, who measures the force $e\mathbf{E}$ on one of his own protons (or anti-protons?) in the arriving pulse of radiation. As a simple example, let **a**(*t*) be zero for $t < 0$, equal to a fixed value a in the positive direction (b) for a short period Δt, and then an equal acceleration in the negative (b) direction for another period Δt, remaining zero thereafter, $t > 2\Delta t$, with the charge at rest. If the distant world of the other physicist is made of matter, he will determine that the initial incoming **E** is in the negative (b) direction. But if his world is an anti-world, his sign convention would have the first force on his anti-proton in the positive (b) direction, and handshakes are out of the question if ever the two physicists should meet. Needless to say, with the large distance **r** between the two physicists, the operational problem would be to achieve enough signal-to-noise to carry out the measurement and sufficiently long-lived physicists to communicate over interstellar distances. It is 8 years round-trip just to and from Alpha Centauri.

In fact, it is apparent from the observed very low level of the expected gamma rays from electron–positron and proton–antiproton annihilation that the universe is filled with matter, to the exclusion of antimatter. Anti-matter is created only as individual anti-particles by collisions of cosmic ray protons, etc., with the nuclei of interstellar matter, and by the creation of electron–positron pairs by gamma rays. So anti-particles are rare, and the behavior of individual anti-particles is very interesting on its own, of course.

With these remarks on positive versus negative charges, consider the point already mentioned that the net electrical charge of the universe is identically zero. One presumes that the equal numbers of electrons and protons were established through creation and annihilation of particles in the initial Big Bang. We may ask, then, what are the experimental and observational upper limits on the net electric charge of the cosmos? Some years ago Lyttleton and Bondi (1959, 1960; Hoyle 1960) considered the consequences of a net electric charge, with the suggestion that a fractional excess charge in the amount of one part in 2×10^{18} would account for the expansion of the universe on the basis of electrostatic repulsion and Newtonian mechanics. It would appear, then, that one part in 2×10^{18} is an upper limit on the positive–negative electric charge asymmetry. For, if the asymmetry were greater, the expansion of the universe would exceed the Hubble constant $R^{-1}dR/dt \approx 80\,\mathrm{km/s}$ per megaparsec inferred from observation. Indeed, with the long-term acceleration of the expansion, presently inferred from observations of distant type Ia supernovae, one might ask if the continuing acceleration could be a consequence of a net electric charge?

Lyttleton and Bondi noted that there may be equal numbers of electrons and protons but the magnitude of the charges may differ slightly, by perhaps one part in 2×10^{18}. Alternatively, the charges may be equal but there is a different total number of electrons and protons. If the former, then the neutron would have a total charge of $\pm e/(2 \times 10^{18})$. If the latter, there would be about 3×10^5 more electrons than protons, or more protons than electrons, in a mole of matter. The mobility of excess electrons might provide electrical conductivity in a cold gas where none exists otherwise. Excess protons represent positive ions of some sort, with possible doping effects.

The Newtonian approach to this electrostatic cosmology considers a sphere of arbitrary radius R at an arbitrary location in an infinite space. The universe outside this sphere is divided into concentric spherical shells extending to unlimited radii. The individual spherical shell contributes neither gravitational nor electrostatic fields to its interior, from which it follows that the cosmos outside the radius R has no effect on the dynamics of R itself. The reader is referred to appendix A for a brief discussion of the mechanics of the sphere with Lagrangian radius $R(t)$. In particular, it follows that the acceleration from electrostatic propulsion is confined to the very early universe, when $R(t)$ was small. So, electrostatic repulsion cannot be responsible for the long-term increase in the Hubble constant.

The Lyttleton-Bondi conjecture challenged experimentalists to look for a slight difference between the magnitudes of the electron and proton charges. For instance, one may contemplate what is implied by the decay of a free neutron into an electron and a proton if the difference in charge is nonzero. One part in 2×10^{18} is a very small difference (see discussion in appendix A), and it was not until Dylla and King (1973) that the experimental upper limit was pushed down to one part in 10^{19}, showing that any difference is negligible so far as cosmology is concerned. Of course, this does not rule out precisely equal charges with a difference in the total number of electrons and protons.

For the present, then, we adopt the simple view that the universe as a whole is electrically neutral, and turn our attention to local neutrality. We are interested in scales of, say, a kiloparsec or less, over which any cosmological net charge has no significant effect. Local electrical neutrality is enforced by the electric field $\mathbf{E}$ ($\nabla \cdot \mathbf{E} = 4\pi\delta$) associated with a net charge density δ. The restoration of charge neutrality ($\delta = 0$) takes place in a characteristic time of the order of the Landau damping time of any plasma oscillations (at the plasma frequency $\omega_p = (4\pi Ne^2/m)^{1/2}$) and of the order of the characteristic resistive damping time $1/4\pi\sigma$, where σ is the electrical conductivity (see appendix B). Thus, there are no surviving electrostatic fields on a macroscopic scale even if somehow a local net charge density were momentarily created. There are, of course, the fluctuating electric fields on the microscopic scale of the Debye radius $(kT/4\pi Ne^2)^{1/2}$, with no implications for the large-scale electric and magnetic fields.

The next question is what happens in an electrically neutral collisionless plasma when a large-scale electric field $\mathbf{E}$ perpendicular to $\mathbf{B}$ is applied by external sources, as one might do in the laboratory? The freely moving electrons and ions are prevented by **B** from streaming in the direction of **E**, which is to say that the conventional concept of large electrical conductivity σ is not applicable. The well-known fact is that the electrons and ions are accelerated by **E** and end up moving around **B** in circles with the appropriate cyclotron frequency, while the circle moves with the steady electric drift velocity $c\mathbf{E} \times \mathbf{B}/B^2$ in the direction perpendicular to both **E** and **B**. In that drifting reference frame there is no electric field (see appendix C) and no further acceleration of the electrons and ions.

So one way or another, there is no significant persistent large-scale electric field in a plasma (collisionless or collision dominated). One might say that a plasma abhors electric fields and invariably finds a means to avoid them. Only by reducing the degree of ionization of a gas to negligible values, e.g., the lower terrestrial atmosphere where we reside, is there a possibility for interesting large-scale electric field effects.

1.4 Electric Charge and Magnetic Field Dominance

In contrast with the electric field, magnetic fields are not erased, because there are no magnetic charges and currents to neutralize them. We have already commented on the absence of magnetic monopoles, and we turn now to the magnetic field of the Galaxy, whose existence places a very low upper limit on the abundance of monopoles (Parker 1970; Turner et al. 1982). If there were n free monopoles per unit volume, each of mass μ, magnetic charge g, and mean conduction velocity $\mathbf{u}$, then the magnetic current density $\mathbf{J}$ would be $gn\mathbf{u}$. The magnetic field of the Galaxy, lying along the spiral arms, has a typical strength $B \approx 4 \times 10^{-6}$ G over a scale Λ of at least one kiloparsec (3×10^{21} cm), indicating magnetic potential differences $\Lambda B \geq 10^{16}$ G cm. The kinetic energy imparted to a monopole by this potential difference implies a monopole velocity of the order of 10^8 cm/s, so we anticipate that the magnetic conduction velocity $\mathbf{u}$ may be a substantial part of the motion of the individual magnetic monopole.

The rate at which the magnetic field **B** does work on **J** is $\mathbf{B} \cdot \mathbf{J} = gn\mathbf{B} \cdot \mathbf{u}$ per unit volume, providing a decline in the magnetic energy density $B^2/8\pi$ at the rate

$$\frac{d}{dt}\frac{B^2}{8\pi} = -gn\mathbf{B} \cdot \mathbf{u} \tag{1.10}$$

The characteristic magnetic dissipation time τ is then

$$\frac{1}{\tau} = -\frac{1}{B}\frac{dB}{dt} = \frac{4\pi ngu}{B} \tag{1.11}$$

That is to say, the conduction flux density nu is related to τ by

$$nu = \frac{B}{4\pi g\tau} \tag{1.12}$$

The continued existence of the galactic magnetic field implies that the dissipation time of the field exceeds the time over which the magnetic field is generated. The magnetic field of the Galaxy exists today, after some 10^{10} years, and it could be argued, therefore, that $\tau > 10^{10}$ years. Put $g = 137e/2$ cgs $= 3.3 \times 10^{-8}$ cgs, with $\tau \geq 3 \times 10^{17}$ s and $B = 4 \times 10^{-6}$ G. The result is $nu < 3 \times 10^{-17}$ monopoles/cm^2 s, i.e., not more than three monopoles intersect a football field (3×10^7 cm^2) in a century, and, of course, perhaps none at all.

If we suppose, on the other hand, that the magnetic field of the Galaxy is regenerated by a galactic dynamo in a time comparable to the rotation period of the Galaxy (2.5×10^8 years or 8×10^{15} s) (Parker 1971a–c,1979), then nu might be 40 times larger, i.e., $nu < 1 \times 10^{-15}$ monopoles/cm^2 s, and the football field might be hit once in a year. (A more detailed analysis can be found in Turner et al. 1982.) It appears that experimental search for monopoles would be a daunting undertaking at best.

Now one might try to avoid the upper limit on monopoles by noting (Turner et al. 1982) that a universe strewn with equal numbers of positive and negative monopoles could experience magnetic monopole Langmuir-type plasma oscillations, with the observed galactic and intergalactic magnetic fields representing the magnetic fields associated with the opposite displacements of positive and negative monopoles. The magnetic monopole plasma frequency would be $\Omega = (4\pi ng^2/\mu)^{1/2}$. The number density n can be estimated from the upper limit on nu, providing an upper limit on Ω and a lower limit on the period of oscillation. If we suppose that $u = 10^{-3}\ c = 300$ km/s, then for $nu < 10^{-15}$ monopoles/cm^2 s, it follows that $n < 3 \times 10^{-23}$/cm^3. The period $2\pi/\Omega$ is then 4×10^7 years. On the other hand, if we suppose that the monopoles have been accelerated by the galactic magnetic field B over a distance $\Lambda = 1$ kpc, the result is $u = 2.3 \times 10^8$ cm/s, so that $n < 4 \times 10^{-24}$/cm^3 and the period of oscillation is 1.1×10^8 years. The monopole oscillations are slow, and one could imagine how the associated magnetic fields would be seen as permanent or static by the transient human observer. However, further investigation shows that the magnetic fields would behave in a curious way, rather contrary to what we seem to see in the cosmos. We imagine the plasma of ionized gas moving with velocity $\mathbf{v}$ relative to the massive background monopole plasma. It is readily shown (Parker 1984, 1987) that the magnetic fields are transported bodily with the velocity **v**/2, i.e., with a velocity halfway between the ionized gas and the monopole plasma. The magnetic field of a galaxy moving relative to the monopole plasma would slip out the trailing side of the galaxy. This seems to be contrary to the observed bulk transport of galactic fields with the motion of the ionized gases contained in the galaxy, and we conclude that the existing large-scale magnetic fields have no association with magnetic monopoles.

The essential point is simply that magnetic fields are not strongly dissipated, whereas large-scale electric fields in the frame of reference of the ionized gases are quickly neutralized. Observations show that nature has not missed this opportunity for proliferation of magnetic fields. The polarization of starlight reddened by passage through interstellar dust, the Faraday rotation effect in radio waves from distant sources, and the synchrotron radiation from energetic electrons show that the plasma filling the universe is everywhere encumbered with magnetic fields. Even solid planets sport magnetic fields, as a consequence of the high electrical conductivity of their rotating convecting liquid interiors. The essential feature for the production and existence of magnetic field is the high electrical conductivity, i.e., the inability to support an electric field in the moving frame of reference of plasma or liquid planetary interior.

This is all so foreign to the situation in the lower terrestrial atmosphere where we reside, the air being an excellent electrical insulator. Here we see none of the magnetic effects, the atmospheric winds blowing freely through the geomagnetic field. Instead, we see such powerful electrostatic phenomena as lightning, driven by potential differences of millions of volts. The tropical thunderstorms charge Earth to some 3–4 $\times 10^5$ V negative with respect to the ionosphere and the space beyond. So there is a downward directed electric field of the order of 1 V/cm here in the lower atmosphere, diminishing upward to the ionosphere at about 100 km altitude. The high density and low temperature of the atmosphere create this unique situation. Indeed, it would appear that the formation of life is possible only in such a situation of low temperature and, hence, negligible electrical conductivity. So, living things can discover the general magnetic character of the cosmos only by remote observation. Only in the physics laboratory can the magnetic plasma conditions be duplicated to some degree.

The fact is that we can understand the remarkable hydrodynamic activity of the astronomical universe only if we have a proper understanding of the dynamical effects of magnetic fields. So, with this in mind, consider the experimental basis for the dynamical theory of electric and magnetic fields.

2 Electric Fields

2.1 Basic Considerations

How does the experimental science of physics connect into the phenomena of the electric field and the magnetic field? How would we recognize an electric field or a magnetic field if we met one? In this chapter we review the conventional formulation of the electric field with special emphasis on the physical reality of the concept, even though our biological senses are not designed to detect either electric or magnetic fields. We cannot see, hear, smell, taste, or feel either field, unless, of course, the field is so strong as to produce secondary mechanical or electrical effects.

For instance, an electric field polarizes our bodies, concentrating one sign of electric charge at one end and the opposite sign at the other. Our hair may rise up, as anyone can attest who has been on a mountain top as a thundercloud passes overhead. Being struck by lightning would be an extreme example. On the other hand, the normal atmospheric electric field of about one volt/centimeter in the high resistivity of the air produces no noticeable electric currents and is below our biological limit of detection.

Strong magnetic fields, of a few kilogauss, encountered, for instance, in the gap of a cyclotron magnet, induce significant electric currents in the body as the human observer moves within the field. The retina of the eye is stimulated to produce tiny scintillations. Electrolysis in the saliva produces an acid flavor in the mouth. These effects suggest that prolonged residence in fields of a few kilogauss or more would interfere with our normal body chemistry. On the other hand, we live in a geomagnetic field of about half a gauss without adverse consequences, and MRI brain scans now expose the living human head to fields of the order of 10^5 G for limited periods. This is a field of biology and medicine that needs serious investigation. It plays a central role in considerations on magnetic shielding of astronauts from exposure to cosmic rays.

So we construct simple instruments to detect electric and magnetic fields under ordinary circumstances. A force (other than gravity) on a charge q tells us of the presence of an electric field. The measured force, denoted by **F**, is put equal to $q\mathbf{E}$, thereby defining the electric field vector **E**. Then, a freely suspended magnetic compass needle is deflected to align itself with the local geomagnetic field **B**, the torque exerted on the needle being $\mathbf{M} \times \mathbf{B}$, where **M** is the magnetic moment of the needle. Thus, the needle tells us of the presence of a magnetic field. Noting that a dielectric needle as well as a metal (electrically conducting) needle are polarized by an electric field, turning so as to align with the electric field, one should really use two freely suspended needles, one magnetized and the other nonmagnetic but dielectric or electrically conducting, to be sure that the response is solely magnetic rather than partly or wholly electrostatic.

We ask the reader to bear with the narrative now, however familiar it maybe, so that we can emphasize a couple of points that are sometimes misunderstood.

2.2 Definition of Charge and Field

The first step is to define the unit of electric charge so that the development can continue on a quantitative basis. For that purpose consider three electric charges, q_1, q_2, q_3, created by rubbing a glass rod with silk, or rubbing cat's fur on hard rubber, or merely by shuffling our shoes across a dry wool rug. The experimental fact of Coulomb's law states that the radial electrostatic repulsion $\mathbf{F}_{12}$ exerted by q_1 on q_2 at a distance r is proportional to q_1q_2/r^2, when r is large compared to the diameter of any of the charges. Following Occam's basic law of theoretical economy, choose the unit of charge so that the proportionality constant is unity, writing

$$\mathbf{F}_{12} = \mathbf{e}_r \frac{q_1 q_2}{r^2} \tag{2.1}$$

where $\mathbf{e}_r$ is the unit vector pointing from q_1 toward q_2. The position of q_2 relative to q_1 is given by the vector $\mathbf{r} = \mathbf{e}_r r$. A repulsive force arises when $q_1q_2 > 0$ and an attractive force when $q_1q_2 < 0$.

Note that the force $\mathbf{F}_{12}$ exerted by q_1 on q_2 is precisely equal and opposite to the force $\mathbf{F}_{21}$ exerted by q_2 on q_1, and both forces are independent of the presence of other charges in the neighborhood. That is to say, the electrostatic force given by (2.1) combines linearly with other electrostatic forces. This linear superposition of electric forces in vacuum is an essential feature of electromagnetic fields, applying to magnetic fields as well.

The unit charge in the cgs electrostatic system produces a force of one dyne when separated by one centimeter from an equal charge. In mks units the unit of charge might reasonably be defined as the charge producing a force of one newton when separated by one meter from an equal charge. The resulting unit of charge would be $10^{9/2}$ times larger than the cgs unit and very small compared to the coulomb (equal to 3×10^9 cgs electrostatic units (esu) of charge). Instead, the construction of the mks, or SI, units was based on the precept that the unit of charge should be the coulomb.

Forcing that point of view introduces a proportionality constant into eqn. (2.1), which was then declared to be the permitivity ε_0 of the vacuum, and the theory became unnecessarily complicated (see section 6.4).

Now to calibrate the three unknown charges q_1, q_2, q_3 in cgs units (esu), one need only measure the electrostatic forces in the three combinations q_1q_2, q_2q_3, q_3q_1 at some standard separation r. This provides three independent measurements, for which it is readily shown that

$$q_1 = \pm\left(\frac{F_{12}F_{31}}{F_{23}}\right)^{1/2} \qquad q_2 = \pm\left(\frac{F_{23}F_{12}}{F_{31}}\right)^{1/2}$$
$$q_3 = \pm\left(\frac{F_{31}F_{23}}{F_{12}}\right)^{1/2} \tag{2.2}$$

where $F_{ij} = q_iq_j/r^2$. The quantities in parentheses are positive no matter what the choice of signs for the charges q_1, q_2, q_3, and the relative signs of the charges are determined experimentally from the signs of the individual forces F_{ij}. There is, of course, an overall sign ambiguity, because the electrostatic forces F_{ij} would be unaffected if the sign of each charge were reversed. As already noted, the convention is to define the charge accumulated on a glass rod rubbed with silk to be positive. That is to say, the electrons rubbed off the glass onto the silk are defined to have a negative charge.

2.3 Concept of Electric Field

Where then do electric fields enter the picture? Experimentally we know that electric charges exert forces on each other, with each element of charge pushing on every other element of charge in the manner described by Coulomb's law, eqn. (2.1). The electric field appears when we ask how it is that the charge q_1 exerts the repulsive force F_{12} on the charge q_2? Noting that the effect is centered around q_1 (which we take to be of very small dimensions—a point charge) and diminishes asymptotically to zero with increasing distance r, we are inclined to the view that q_1 is surrounded by a "sphere of influence," whatever that may be. Define the electric field of q_1 to be the force per unit charge exerted on q_2, so that $\mathbf{F}_{12} = q_2\mathbf{E}$. It follows that the electric field $\mathbf{E}$ at q_2 is given by

$$\mathbf{E}(\mathbf{r}) = \mathbf{e}_r\frac{q_1}{r^2} \tag{2.3}$$

where again $\mathbf{e}_r$ is the unit vector pointing radially away from q_1. Thus, quite generally, the forces exerted by q_1 on q_2 and q_3 are given by $\mathbf{F}_{12} = q_2\mathbf{E}(\mathbf{r}_{12})$

and $\mathbf{F}_{13} = q_3\mathbf{E}(\mathbf{r}_{13})$, respectively, where $\mathbf{r}_{12}$ represents the location of q_2 relative to q_1, and $\mathbf{r}_{13}$ the location of q_3 relative to q_1.

The essential point here is that the measured force $\mathbf{F} = q\mathbf{E}$ on a charge q defines the electric field $\mathbf{E}$ at the position of the charge q in the frame of reference moving with the charge itself. That is to say, if an electric charge q experiences a force (apart from gravity), then there is an electric field in the frame of reference moving with q. There are other electric fields, of course, in other moving frames of reference, but they apply only to charges at rest in those respective reference frames. This seeming trivial point is not always appreciated, and the reader is referred to Vasyliunas and Song (2005) for a discussion of the problem in a partially ionized gas.

Now the electric field associated with a broad distribution of charge, with density $\delta(\mathbf{r})$, follows as the linear superposition of the electric fields of all of the infinitesimal elements $\delta(\mathbf{r})d^3\mathbf{r}$ of charge, providing the relation

$$\mathbf{E}(\mathbf{r}) = \int_V \frac{d^3\mathbf{r}'\delta(\mathbf{r}')(\mathbf{r} - \mathbf{r}')}{|\mathbf{r} - \mathbf{r}'|^3} \tag{2.4}$$

where the integration over $\mathbf{r}'$ is carried out over the volume V wherein $\delta(\mathbf{r})$ is nonvanishing.

The total electric flux $\Phi = \int d\mathbf{S} \cdot \mathbf{E}$ outward across a closed surface S enclosing a point charge q is readily computed. Put the origin of the coordinates at the charge and let r represent distance from the origin. Referring to Fig. 2.1 consider the electric field flux in the infinitesimal solid angle $d\Omega$ from the point charge q. With the element of area $dS = r^2d\Omega$ subtended by $d\Omega$, it follows from eqn. (2.3) that the electric flux out through $d\Omega$ is

$$d\Phi = q d\Omega \tag{2.5}$$

Hence, over the entire 4π steradians,

$$\Phi = 4\pi q \tag{2.6}$$

no matter what the shape and size of the enclosing surface S. The fact that Φ is independent of the size of the enclosing surface indicates conservation of electric flux throughout the space between charges. That is to say,

$$\nabla \cdot \mathbf{E} = 0 \tag{2.7}$$

This can also be shown by computing $\nabla \cdot \mathbf{E}$ from eqn. (2.3) for the field from each basic element of charge q or dq and applying the principle of

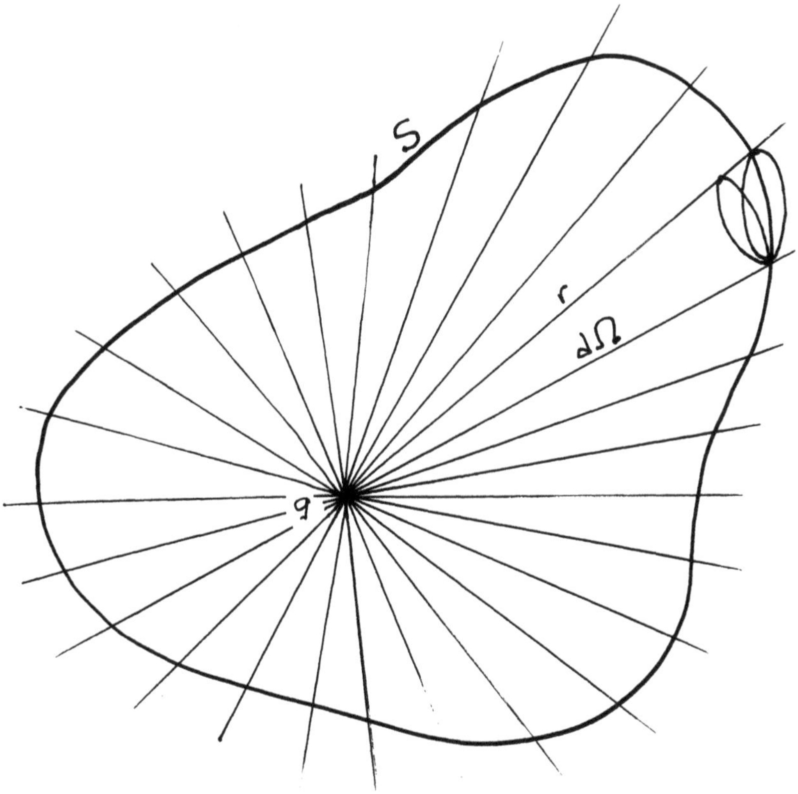

Fig. 2.1 A sketch of the radial electric field from a point charge q, showing the elemental solid angle $d\Omega$ and the arbitrary enclosing surface S.

linear superposition of electric fields. The total flux outward through S depends only on the total charge contained within S.

Note that if $\delta(\mathbf{r})$ represents the charge density throughout the interior of S, then we can write

$$\Phi = 4\pi \int_S dV \delta(\mathbf{r}) \tag{2.8}$$

The definition of Φ as the integral of $d\mathbf{S} \cdot \mathbf{E}$ over the surface S can be rewritten as

$$\Phi = \int_S dV \nabla \cdot \mathbf{E} \tag{2.9}$$

with the aid of Gauss's theorem. These two expressions for Φ apply to all surfaces S. Hence, the integrands must be equal at every point, and

$$\nabla \cdot \mathbf{E} = 4\pi\delta \tag{2.10}$$

everywhere, for if there were a region where this were not correct, enclosing that region with a surface S would lead to a contradiction.

2.4 Physical Reality of Electric Field

Having remarked on the properties of the electric field **E** associated with electric charge, the obvious question is whether **E** is anything more than a convenient mathematical construction, or whether, in addition, there is really something there? The answer is, of course, that the electric field is a condition in space that represents a local energy density, and hence a local mass density, and the electric field transmits stresses across space from one region to another. Indeed, we began the discussion with the stresses that the electric field transmits from one electric charge to another. What is more, an electric field in one reference frame implies an electric field in almost every other relatively moving reference frame.

To explore the physical properties of the electric field, consider some simple displacements of the electric charges that are the source of **E**. Work is done in moving the charges, and the principle of conservation of energy requires that the energy represented by the work should reside somewhere in the physical universe. So where could the energy be except in the newly created electric field?

Consider, then, two parallel broad sheets (width a and length b) of electrical insulating material, one with uniform surface charge density $+\sigma$ and the other with surface charge density $-\sigma$, and separated by a distance h. The field between the two sheets is $E = 4\pi\sigma$, given that the total electric flux from unit area is 4π times the charge on that area. The field of the sheet with charge density σ would by itself be an outward field of strength $2\pi\sigma$ in both directions from the sheet, sketched in Fig. 2.2. Superposing the fields of both sheets gives the field of $4\pi\sigma$ in the space between the sheets, whereas outside that space the two fields cancel and there is nothing. Now each sheet resides in the field of the other, from which it follows that each sheet is attracted toward the other with a force $2\pi\sigma \times \sigma$ per unit area, which is equal to $E^2/8\pi$. This indicates that there is a net tension $E^2/8\pi$ transmitted along the field.

Suppose now that the distance between the two sheets is increased by some small amount Δh, in opposition to the force $E^2/8\pi$ pulling them toward each other. The volume of field is thereby increased by $\Delta h/\text{cm}^2$

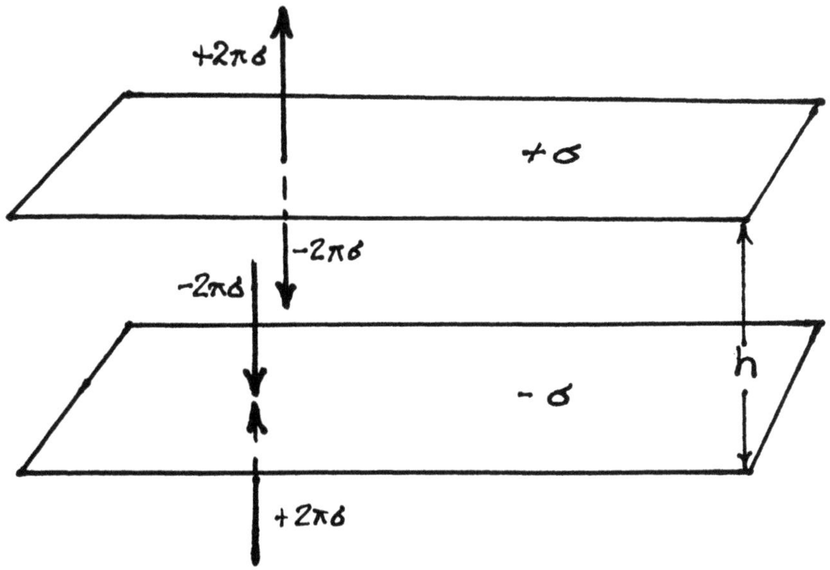

Fig. 2.2 A sketch of the electric field $E = 2\pi\sigma$ directed outward in both directions from two plane parallel sheets with opposite surface charge densities $\pm\sigma$ and separated by a distance h.

and the work done on the system is $\Delta h(E^2/8\pi)/\text{cm}^2$. That input of energy evidently resides in the newly created electric field in the volume Δh, from which it follows that the energy density U of the field is

$$U = \frac{E^2}{8\pi} \tag{2.11}$$

If the separation of the sheets is then decreased to h again, this field energy is converted back into the work that the attractive force does on the two sheets. That is to say, the field behaves exactly as if it had this energy density, and, as already emphasized, reality is made up of the manner in which things behave experimentally.

Now if the field has energy density $E^2/8\pi$, it follows that the field has a mass density $E^2/8\pi c^2$. It follows that an electric field has inertia and is the source of gravitational field. We will find in the next chapter that the magnetic field **B** has the equivalent mass density $B^2/8\pi c^2$. This energy and mass equivalence of electric and magnetic fields should come as no surprise, of course, because electromagnetic radiation, e.g., sunlight, consists only of electric and magnetic field and certainly carries energy. So the energy of electric and magnetic fields is a dazzling reality. Later in the conversation we note Poynting's formal derivation of the electromagnetic energy density and stress tensor from Maxwell's equations.

As a final comment on the physical reality of the electric field, note that in the hypothetical magnetic monopole universe, mentioned briefly at the end of section 1.2, the magnetic monopole plasma filling that universe would be tied to the large-scale electric fields, just as the magnetic field and plasma are tied together in our own universe. So the electric field would be a very real constituent of a ponderable fluid medium, representing stress and energy transported bodily by the bulk motion of the monopole plasma.

2.5 Electric Field Pressure

Suppose that the separation of the two parallel and oppositely charged sheets is held fixed at the value h, while the sheets are stretched (uniformly dilated) in the direction of their length b by some small amount Δb, sketched in Fig. 2.3. The total charge $\pm ab\sigma$ on each sheet is conserved, so that the charge density on each sheet is diminished by some small amount $\Delta\sigma$. Charge conservation requires that

$$a(b + \Delta b)(\sigma - \Delta\sigma) = ab\sigma$$

Hence, for $\Delta b \ll b$,

$$\Delta\sigma = \sigma\frac{\Delta b}{b} \tag{2.12}$$

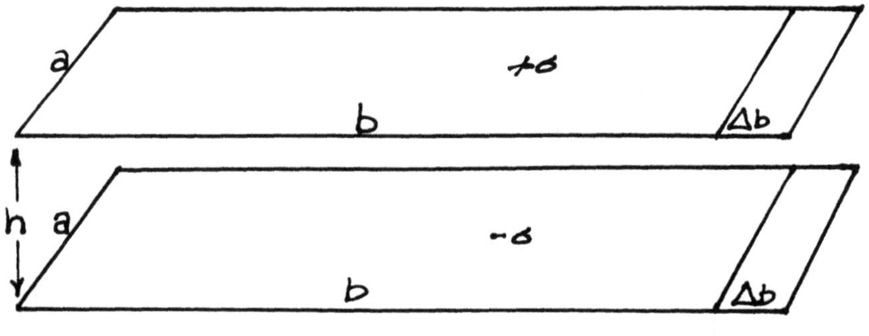

Fig. 2.3 A sketch of the uniform dilatation of the two oppositely charged parallel sheets, extending their length from b to $b + \Delta b$ while conserving total charge.

The electric field between the two sheets is diminished to $E - \Delta E$, where $\Delta E = 4\pi\Delta\sigma$. The energy density is

$$U - \Delta U = \frac{(E - \Delta E)^2}{8\pi}$$

from which it follows that

$$\begin{aligned} \Delta U &= \frac{E\Delta E}{4\pi} \\ &= \frac{E^2}{4\pi}\frac{\Delta b}{b} \end{aligned} \tag{2.13}$$

The total field energy is

$$a(b + \Delta b)(U - \Delta U) = a\frac{E^2}{8\pi}(b - \Delta b)$$

diminished by $a\Delta b E^2/8\pi$ from the original $abE^2/8\pi$. The energy of **E** has evidently gone into doing work on the mechanical system during the increase Δb in the length. Evidently, then, the field exerts a transverse pressure P given by

$$P = \frac{E^2}{8\pi} \tag{2.14}$$

so that P does work in the amount $a\Delta bP = \Delta U$.

In summary, the stresses in the electric field **E** are equivalent to an isotropic pressure $E^2/8\pi$ and a tension $E^2/4\pi$ along the field, so that there is a net tension $E^2/8\pi$ along the field, as noted in section 2.4. The net effect is described by the stress tensor

$$M_{ij} = -\delta_{ij}\frac{E^2}{8\pi} + \frac{E_iE_j}{4\pi} \tag{2.15}$$

using the convention that pressure is negative and tension is positive.

One may ask why the net tension $E^2/8\pi$ along the field is interpreted as the sum of a pressure $-E^2/8\pi$ and a tension $E^2/4\pi$. The answer is that a pressure $E^2/8\pi$ in the two dimensions perpendicular to the field and only a tension $E^2/8\pi$ in the direction parallel to the field allows no tensor representation M_{ij}. It is then not possible to construct covariant equations for such things as the equilibrium of an electric field. Consider, for instance, the static equilibrium of E_i at a point Q under the stresses in the direction parallel to the field E_i. Let s represent distance measured along the field line through Q. The field magnitude is equal to the field component E_s, while the perpendicular field component vanishes. Equilibrium requires that $\partial M_{ij}/\partial x_j = 0$, or

$$-\frac{\partial}{\partial x_i}\frac{E^2}{8\pi} + \frac{E_j}{4\pi}\frac{\partial E_i}{\partial x_j} = 0$$

upon using the divergence condition, $\partial E_j/\partial x_j = 0$, in empty space. For the component along the field, $i = s$, this reduces to

$$-\frac{\partial}{\partial s}\frac{E_s^2}{8\pi} + \frac{E_s}{4\pi}\frac{\partial E_s}{\partial s} = 0$$

which is a trivial identity. On the other hand, if there were nothing but a tension $E^2/8\pi$ along the field, one would be inclined to some such condition as $\partial/\partial s(E^2/8\pi) = 0$ for equilibrium. That would allow no equilibrium except in a uniform field, contrary to experimental fact.

3 Magnetic Fields

3.1 Basic Considerations

The magnetic field **B** bears the same relation to a hypothetical magnetic charge m as the electric field bears to the electric charge q, so the development of the previous chapter can be taken over completely. The stress tensor for a magnetic field B_i is identical in form to that for the electric field. The energy density is obviously $B^2/8\pi$ and the magnetic field is a physical reality just like the electric field. Electric and magnetic fields superpose linearly, so that the complete electromagnetic stress tensor is

$$M_{ij} = -\delta_{ij}\frac{E^2}{8\pi} + \frac{E_iE_j}{4\pi} - \delta_{ij}\frac{B^2}{8\pi} + \frac{B_iB_j}{4\pi} \tag{3.1}$$

and the total energy density is

$$U = \frac{E^2 + B^2}{8\pi}$$

As remarked earlier, it appears that there are few, if any, magnetic charges in the universe, so that we have the universal condition that

$$\nabla \cdot \mathbf{B} = 0$$

This is perhaps a convenient place to remark that a magnetic field exerts no force whatever on an electric charge. Only an electric field in the frame of reference of the charged particle exerts force on the particle.

In the absence of magnetic charges, magnetic fields appear only in association with electric currents and in association with time varying electric fields, in the manner described by eqn. (1.6). In the laboratory we create static magnetic fields by driving an electric current through a coil of wire. The emf driving the current is the source of energy that creates the magnetic field, so the emf and the current are clearly the *cause* of the magnetic field. On the other hand, in the cosmos the deformation of the magnetic field embedded in the swirling plasma *causes* the flow of electric current in the plasma in the manner described by eqn. (1.6), because the energy that drives the current comes from the magnetic field. That is to say, $4\pi\mathbf{j}$ is maintained close to the value $c\nabla \times \mathbf{B}$ by the fact that any deviation produces a $\partial\mathbf{E}/\partial t$ that quickly provides an **E** that drives j to the required value. So in the cosmos the large-scale currents are obliged by $\partial\mathbf{E}/\partial t$ to conform to Ampere's law,

$$4\pi\mathbf{j} = c\nabla \times \mathbf{B} \tag{3.2}$$

In view of the small but nonvanishing friction between the relative motions of the electrons and ions, there is a continuing trickle of energy from the field to the current to maintain the flow of current required by Ampere, from which it follows that the field is the continuing cause of the current and not vice versa.

The curious popular notion that the electric current *causes* the magnetic fields in the cosmos has led to the even more curious notion that the electric current is the more fundamental dynamical variable. Then, since currents are driven by electric fields, it is declared that the fundamental dynamical variables are **E** and **j**. As already noted, the difficulty is that there are no tractable dynamical equations for **E** and **j**. The current is dynamically passive, consisting of no more than the tiny inertia of the electron conduction velocity, while, as we shall see, the stresses in the electric field are small to second order in v/c and quite negligible. The dynamics of the plasma–magnetic field system is driven by the magnetic stress and the inertia and pressure of the plasma.

3.2 Experimental Connection

The reality of our universe is that magnetic charges, convenient for describing the close analogy between the properties of electric and magnetic fields, do not occur in the laboratory. So we consider here the much more complicated, and rather more interesting, route by which the physics of magnetic fields was developed from laboratory experiments.

The connection between magnetic fields and electric currents began with Oersted in 1819, using a compass needle in the neighborhood of a wire carrying a current I. The compass needle was aligned by the familiar magnetic field of Earth until the current was switched on. At that point the effect of the current took over and the reorientation of the compass established the existence of a magnetic field in association with an electric current. The compass needle provided a means to map the form of the magnetic field associated with the current. This experimental revelation led to active experimental pursuit of the magnetic field–electric current connection, and in 1820 Biot pointed out that the magnetic field

$d\mathbf{B}(\mathbf{r})$ at the position $\mathbf{r}$ produced by an infinitesimal segment of current $d\mathbf{I}(\mathbf{r}') = I(\mathbf{r}')ds$ at the position $\mathbf{r}'$ is given by

$$d\mathbf{B}(\mathbf{r}) = K\frac{d\mathbf{I}(\mathbf{r}') \times (\mathbf{r} - \mathbf{r}')}{|\mathbf{r} - \mathbf{r}'|^3} \tag{3.3}$$

Subsequently Biot, Savart, and Ampere performed a variety of laboratory measurements, testing this current–field relation and determining the proportionality constant K. It turns out that in cgs units K is equal to $1/c$, where c is the speed of light. Thus, with $K = 1/c$ eqn. (3.3) may be used to define the unit of magnetic field in terms of the unit of current.

For a continuous distribution of the current density $\mathbf{j}(\mathbf{r}')$ the current segment associated with the volume element $d^3\mathbf{r}'$ is

$$d\mathbf{I}(\mathbf{r}') = \mathbf{j}(\mathbf{r}')d^3\mathbf{r}' \tag{3.4}$$

so that linear superposition of the individual contributions of eqn. (3.3) yields

$$\mathbf{B}(\mathbf{r}) = \frac{1}{c}\int d^3\mathbf{r}'\frac{\mathbf{j}(\mathbf{r}') \times (\mathbf{r} - \mathbf{r}')}{|\mathbf{r} - \mathbf{r}'|^3} \tag{3.5}$$

known as the Biot-Savart integral, providing the magnetic field for a prescribed electric current distribution.

3.3 Differential Form of Ampere's Law

The differential form of Ampere's law, providing the electric current for a prescribed magnetic field, is readily computed from the curl of the Biot-Savart integral,

$$\begin{aligned}\nabla \times \mathbf{B}(\mathbf{r}) &= \frac{1}{c}\int d^3\mathbf{r}'\nabla \times \left[\mathbf{j}(\mathbf{r}') \times \frac{\mathbf{r} - \mathbf{r}'}{|\mathbf{r} - \mathbf{r}'|^3}\right] \\ &= \frac{1}{c}\int d^3\mathbf{r}'\left\{\left(\frac{\mathbf{r} - \mathbf{r}'}{|\mathbf{r} - \mathbf{r}'|^3}\cdot\nabla\right)\mathbf{j}(\mathbf{r}') - [\mathbf{j}(\mathbf{r}')\cdot\nabla]\frac{\mathbf{r} - \mathbf{r}'}{|\mathbf{r} - \mathbf{r}'|^3}\right. \\ &\left. + \mathbf{j}(\mathbf{r}')\nabla\cdot\frac{\mathbf{r} - \mathbf{r}'}{|\mathbf{r} - \mathbf{r}'|^3} - \frac{\mathbf{r} - \mathbf{r}'}{|\mathbf{r} - \mathbf{r}'|^3}\nabla\cdot\mathbf{j}(\mathbf{r}')\right\}\end{aligned} \tag{3.6}$$

where we have used the well-known vector identity for the curl of a vector product. The gradient operator ∇ applies only to the point $\mathbf{r}$, while the current is a function only of $\mathbf{r}'$. Hence, the first and fourth terms vanish, reducing (3.6) to

$$\nabla \times \mathbf{B}(\mathbf{r}) = \frac{1}{c}\int d^3\mathbf{r}'\left\{\mathbf{j}(\mathbf{r}')\nabla\cdot\frac{\mathbf{r}-\mathbf{r}'}{|\mathbf{r}-\mathbf{r}'|^3} - [\mathbf{j}(\mathbf{r}')\cdot\nabla]\frac{\mathbf{r}-\mathbf{r}'}{|\mathbf{r}-\mathbf{r}'|^3}\right\} \tag{3.7}$$

It must be understood that the volume over which the integral is evaluated contains all the currents, with no volume omitted in which $\mathbf{j}(\mathbf{r}') \neq 0$. It follows that $\mathbf{j}(\mathbf{r}')$ vanishes at the surface of the volume.

Now

$$\frac{\mathbf{r}-\mathbf{r}'}{|\mathbf{r}-\mathbf{r}'|^3} = -\nabla\left(\frac{1}{|\mathbf{r}-\mathbf{r}'|}\right) \tag{3.8}$$

So the first term in the integrand of (3.7) can be written $-\mathbf{j}(\mathbf{r}')\nabla^2|\mathbf{r}-\mathbf{r}'|^{-1}$, which is equal to $+4\pi\mathbf{j}(\mathbf{r}')\delta(\mathbf{r}-\mathbf{r}')$, and $\delta(\mathbf{r}-\mathbf{r}')$ is the Dirac delta function. Hence, the integration over $\mathbf{r}'$ converts the first term to $4\pi\mathbf{j}(\mathbf{r})$.

The second term in the integrand in eqn. (3.7) can be transformed by noting that

$$\nabla\left(\frac{\mathbf{r}-\mathbf{r}'}{|\mathbf{r}-\mathbf{r}'|^3}\right) = -\nabla'\left(\frac{\mathbf{r}-\mathbf{r}'}{|\mathbf{r}-\mathbf{r}'|^3}\right) \tag{3.9}$$

where ∇' represents the gradient in $\mathbf{r}'$. Then, adopting index notation, it follows that

$$[\mathbf{j}(\mathbf{r}')\cdot\nabla]\frac{\mathbf{r}-\mathbf{r}'}{|\mathbf{r}-\mathbf{r}'|^3} = -\frac{\partial}{\partial x'_k}\left[\mathbf{j}_k(\mathbf{r}')\frac{\mathbf{r}-\mathbf{r}'}{|\mathbf{r}-\mathbf{r}'|^3}\right] + \frac{\mathbf{r}-\mathbf{r}'}{|\mathbf{r}-\mathbf{r}'|^3}\frac{\partial \mathbf{j}_k(\mathbf{r}')}{\partial x'_k} \tag{3.10}$$

The first term on the right-hand side is a divergence, and, upon application of Gauss's theorem, provides an integral over the surface of the volume. But the current vanishes on the surface, as already noted, so the term contributes nothing. The second term vanishes under steady conditions of current conservation, $\nabla \cdot \mathbf{j}(\mathbf{r}) = 0$. Thus, eqn. (3.7) reduces to

$$\nabla \times \mathbf{B}(\mathbf{r}) = 4\pi\mathbf{j}(\mathbf{r}) \tag{3.11}$$

This is the familiar differential form of Ampere's law.

The differential form of Ampere's law compliments the Biot-Savart integral form. We note here the third form, in terms of the total current through a surface S enclosed by a closed contour Λ. The line integral of $\mathbf{B}$ around the closed contour can be transformed, using Stokes's theorem and eqn. (3.11), into the total current I flowing across the surface S. We have

$$\oint_{\Lambda} d\mathbf{s} \cdot \mathbf{B} = \int_{S} d\mathbf{S} \cdot \nabla \times \mathbf{B}$$
$$= \frac{4\pi}{c} \int_{S} d\mathbf{S} \cdot \mathbf{j}$$
$$= \frac{4\pi}{c} I \tag{3.12}$$

It follows that the line integral of $\mathbf{B}$ around any closed contour is equal to $4\pi/c$ times the total current I flowing through the enclosed area. Thus, for instance, a twisted flux bundle of finite radius carries no net current, because if the magnetic field of the bundle is confined to a finite radius, then beyond that radius the line integral of the field around the bundle vanishes.

3.4 Energy and Stress

It is now a straightforward exercise to work out the energy density, pressure, and tension of a magnetic field for a quasi-steady magnetic field associated with an electric current in the manner described by Ampere's law. We know the results already, based on the analogy with the electric field associated with electric charges and described by eqn. (3.1). So we shall be brief.

Consider the circular magnetic field between two concentric circular cylinders of radius a and b ($a < b$) and length L ($>>a, b$), with an electric current I flowing along the inner cylinder and the return current $-I$ on the outer cylinder. The cylinders are perfectly conducting, and there are perfectly conducting end plates to transfer I from one cylinder to the other. Apply Ampere's law in the form of eqn. (3.12) to the circular field lines ϖ = *constant*, where ϖ is the radial distance from the axis of the cylinders. It follows from eqn. (3.12) that $2\pi\varpi B(\varpi) = 4\pi I/c$, so that

$$B(\varpi) = \frac{2I}{c\varpi} \tag{3.13}$$

The total magnetic flux Φ circling the inner cylinder, $\varpi = a$, over the entire length L is

$$\Phi = L\int_a^b d\varpi B(\varpi)$$

$$= \frac{2IL}{c}\ln\frac{b}{a} \tag{3.14}$$

Consider, then, the energy in the magnetic field. Suppose that $I = 0$ initially. At time $t = 0$ an emf $V(t)$ is applied and $I(t)$ increases from zero. The total energy input $W(t)$ after a time t is given by

$$W(t) = \int_0^t dt'\, V(t')I(t')$$

Faraday's experimental law of induction (see chapter 6) states that $V(t) = +(1/c)d\Phi/dt$, so that with the aid eqn. (3.14), we have

$$W(t) = \frac{1}{2L\ \ln b/a}\int_0^t dt'\,\Phi(t')\frac{d\Phi}{dt'}$$

$$= \frac{\Phi^2}{4L\ \ln b/a} \tag{3.15}$$

Conservation of energy implies that the energy input goes into creating the magnetic field, so that the total magnetic field energy $Z(t)$ is equal to $W(t)$. Use eqn. (3.14) to express I in terms of Φ and write eqn. (3.13) as

$$B(\varpi) = \frac{\Phi(t)}{L\ln(b/a)\varpi} \tag{3.16}$$

The fact that $W(t)$ is quadratic in $\Phi(t)$ indicates that the energy density of the magnetic field is quadratic in $B(\varpi)$. So assume that the energy density can be written as $QB(\varpi)^2$, the constant Q to be determined. Then,

$$Z(t) = 2\pi L\int_a^b d\varpi\varpi QB^2(\varpi)$$

$$= \frac{2\pi Q\Phi(t)^2}{L\ln(b/a)}$$

When this is compared with eq. (3.15) it is evident that $Q = 1/8\pi$, and the energy density U of a magnetic field is the expected

$$U = \frac{B^2}{8\pi} \tag{3.17}$$

Suppose, then, that the applied emf is reduced to zero, leaving the current flowing freely and steadily along the two cylinders, with the field described by eqn. (3.16). By mechanical means increase the radius b of the outer cylinder to $b + \Delta b$. The total magnetic flux Φ does not change, of course, because if it did, there would be an emf induced into the path of the freely flowing electric current. That path is through an infinitely conducting material, which guarantees that there can be no emf around that path. With constant Φ, it follows that the magnetic energy

$$Z = \frac{\Phi^2}{4L\ln b/a} \tag{3.18}$$

changes by

$$\Delta Z = -\frac{\Phi^2 \Delta b}{4Lb\ln^2 b/a} \tag{3.19}$$

Presumably, this decrease in field energy is a consequence of the work done by the field on the receding cylinder as b increases to $b + \Delta b$. Denoting the magnetic pressure on $\varpi = b$ by P, it follows that the work done is $\Delta W = 2\pi bLP\Delta b$. Conservation of energy requires that $\Delta Z + \Delta W = 0$, from which it follows that the magnetic pressure on the expanding cylinder $\varpi = b$ is

$$P = \frac{\Phi^2}{8\pi Lb^2\ln^2 b/a}$$

$$= \frac{B(b)^2}{8\pi} \tag{3.20}$$

as expected. The same result is obtained if we vary the inner radius a or the length L.

The tension in the magnetic field is readily deduced from the equilibrium between the tension along the field around the radius ϖ and the radial gradient of the magnetic pressure $B(\varpi)^2/8\pi$. If we denote the tension force per unit area by $\Theta(\varpi)$, then the inward force per unit volume is $\Theta(\varpi)/\varpi$. Balancing this against the outward pressure gradient gives

$$\begin{aligned}\frac{\Theta}{\varpi} &= -\frac{d}{d\varpi}\frac{B(\varpi)^2}{8\pi} \\ &= -\frac{I^2}{2\pi c^2}\frac{d}{d\varpi}\left(\frac{1}{\varpi^2}\right) \\ &= +\frac{I^2}{\pi c^2 \varpi^3}\end{aligned}$$

Hence,

$$\begin{aligned}\Theta &= \frac{I^2}{\pi c^2 \varpi^2} \\ &= \frac{B^2}{4\pi} \qquad (3.21)\end{aligned}$$

as expected. So, as already suggested from the analogy with the electric field stresses, the Maxwell stress tensor for the magnetic field is

$$M_{ij} = -\delta_{ij}\frac{B^2}{8\pi} + \frac{B_i B_j}{4\pi} \qquad (3.22)$$

As with the electric field, the magnetic field is an energy-bearing stress system. It is a real physical entity that contributes to the inertial and gravitational mass of the system. To assume otherwise would lead to the remarkable complication that the energy $W(t)$ going into the creation of the magnetic field would represent a net loss to the effective mass of the system.

Another way to look at the physical reality of the magnetic field is through MHD, wherein the magnetic field is inextricably tied to the swirling plasma, providing the fluid plasma with an internal stress system and internal energy that moves along with the plasma.

3.5 Detecting a Magnetic Field

Let us go back to the basic question of detecting and measuring a magnetic field, passed over so quickly in sections 2.1 and 3.2. It was pointed out that a magnetic field is readily detected and measured by introducing a small freely swinging magnetic compass needle, following Oersted's example of 1819. The needle aligns itself with **B**, and the torque producing the alignment is $\mathbf{M} \times \mathbf{B}$ for a needle with magnetic moment **M**.

Note that the simultaneous presence of an electric field **E** would induced an electric dipole moment in the needle, and, hence, an additional torque. So it is simplest in principle to place the magnetic needle in a nonmagnetic Faraday cage to shield the needle from the ambient **E**.

Now we determined the electric field **E** in the laboratory frame of reference by measuring the force **F** on a charge q at rest in the laboratory reference frame, giving $\mathbf{E} = \mathbf{F}/q$ (sections 2.1 and 3.2). To determine the magnetic field in the laboratory with a charge q, we move q with velocity **v** so that q experiences the electric field $\mathbf{E}'$ in another frame of reference. The nonrelativistic Lorentz transformation gives $\mathbf{E}'$ as

$$\mathbf{E}' = \mathbf{E} + \frac{\mathbf{v} \times \mathbf{B}}{c}$$

so that the change $\Delta\mathbf{E}$ in electric field at q is

$$\Delta\mathbf{E} = \frac{\mathbf{v} \times \mathbf{B}}{c}$$

and the force on q changes by

$$\begin{aligned} \Delta\mathbf{F} &= q\Delta\mathbf{E} \\ &= \frac{q}{c}\mathbf{v} \times \mathbf{B} \end{aligned}$$

It follows that the component of **B** perpendicular to **v** is

$$\mathbf{B}_\perp = \frac{c}{qv^2}\Delta\mathbf{F} \times \mathbf{v}$$

Then measure the force $\Delta\mathbf{F}_u$ on q when q has the velocity

$$\mathbf{u} = v\frac{\mathbf{B}_\perp}{B_\perp}$$

relative to the laboratory. With

$$\Delta\mathbf{F}_u = \frac{q}{c}\mathbf{u} \times \mathbf{B}$$

$$= \frac{qv}{cB_\perp} \mathbf{B}_\perp \times \mathbf{B}$$

it follows that the field component $\mathbf{B}_\parallel (= \mathbf{B} - \mathbf{B}_\perp)$ parallel to **v** is given by

$$\mathbf{B}_\parallel = \frac{c}{qvB_\perp} \Delta\mathbf{F}_u \times \mathbf{B}_\perp$$

Another approach would be to measure the force on q for a variety of velocities **v**, determining the direction of **v** for which $\Delta\mathbf{F} = 0$. That is the direction of **B**, of course, so that any velocity **w** perpendicular to that direction provides the force

$$\Delta\mathbf{F} = \frac{q}{c} \mathbf{w} \times \mathbf{B}$$

from which it follows that

$$\mathbf{B} = \frac{c}{qw^2} \Delta\mathbf{F} \times \mathbf{w}$$

Needless to say, there are more convenient instruments for measuring **B**, based on the Zeeman effect and other atomic processes. We have been concerned here only with the basic principles for direct detection and measurement of **B**.

No discussion of the detection of magnetic field would be complete without mentioning the Aharonov-Bohm effect (Ehrenberg and Siday 1949; Aharonov and Bohm 1959; Wu and Yang 1975; Olariu and Popescu 1985), which arises through the "magic" of quantum mechanics. The effect provides a direct measurement of the total magnetic flux through a designated area based on the difference in path integrals around the area on either side. Specifically, the phase of the electron wave function varies as exp $i2\pi S/h$, where

$$\frac{S}{h} = \frac{e}{h} \int \left(dt\varphi - \frac{d\mathbf{s} \cdot \mathbf{A}}{c} \right)$$

in terms of the action integral along the path of the wave. Here $\varphi(\mathbf{r}, t)$ is the electric potential and $\mathbf{A}(\mathbf{r}, t)$ is the vector potential, with $\mathbf{B}(r, t) = \nabla \times \mathbf{A}$. The conjugate momentum is $\mathbf{p} - e\mathbf{A}/c$ and the Hamiltonian is

$$H = \frac{(\mathbf{p} - e\mathbf{A}/c)^2}{2m} + e\varphi$$

Now imagine a long, thin, finely wound, current carrying solenoid containing a total magnetic flux Φ. The open ends of the solenoid are so far away that the field outside the solenoid falls to zero everywhere along the middle length of the solenoid. Then, if $d\mathbf{S}$ denotes an element of area in the cross section of the solenoid, it follows that

$$\Phi = \int d\mathbf{S} \cdot \mathbf{B}$$

$$= \oint d\mathbf{s} \cdot \mathbf{A}$$

upon writing $\mathbf{B} = \Delta \times \mathbf{A}$ and then applying Stokes's theorem, where the integral is around the periphery of the cross section of the solenoid.

In the absence of an electric field ($\varphi = 0$), it follows that the total phase contribution around the enclosed magnetic field is

$$\frac{S}{h} = -\frac{e\Phi}{hc}$$

Then shield the solenoid so that a monoenergetic electron beam directed against the solenoid does not penetrate the region of magnetic field but passes by freely on both sides of the solenoid. That is to say, the electron wave function is zero within the solenoid, and the shielded solenoid produces a shadow of width d in the electron beam, where d is the outside diameter of the electron shield enclosing the solenoid.

At a large distance L downstream from the solenoid the split electron beam impacts a phosphorescent screen. Diffraction of the electron beam around the opaque solenoid produces interference fringes, of course, with the angular separation $\Delta\theta$ of neighboring bands given by

$$\Delta\theta = \frac{\lambda}{d}$$

in terms of the de Broglie wavelength $\lambda = h/p$. The linear separation of the fringes is then $L\Delta\theta = L\lambda/d$.

The presence of the magnetic flux Φ enclosed in the solenoid introduces an additional phase difference between the two interfering electron beams amounting to $2\pi S/h$, which can be varied by controlling the electric current in the solenoid. The number of fringes displaced by this phase difference is S/h.

The essential point is that the electron fringe system is displaced by the presence of a magnetic flux Φ between the two halves of the beam, even though the magnetic field is not encountered by the electrons. There is no magnetic field in the volume of space swept out by the electron beam. The presence of the magnetic field in the solenoid is made known to the electron beam through the vector potential **A**, providing the relative difference in the phase of the wave functions passing on opposite sides of the solenoid.

In classical (nonquantum mechanical) physics, which is the subject of these *Conversations*, the vector potential **A** is only a mathematically convenient way to express a magnetic field **B** ($= \nabla \times \mathbf{A}$), guaranteeing for any **A** that $\nabla \cdot \mathbf{B} = 0$. The vector potential has no physical presence, as distinct from the electric field **E** and the magnetic field **B**, which have stress and energy at every point. The physical Aharonov-Bohm effect is, then, the quantum mechanical manifestation of the vector potential, however immaterial the vector potential may be in the classical world. The electrons never see the magnetic field itself, but the total flux of magnetic field shows up in the fringe shift.

Aharonov and Bohm noted a similar effect achieved by a short pulse of beamed electrons split in two and passing separately through two long thin parallel Faraday cages (metal tubes) whose electrical potentials differ by $\Delta\varphi$. The phase difference between the two emerging beams of electrons is $(2\pi e/h)\int dt \Delta\varphi(t)$. It must be recognized that the two halves of the electron beam experience electric fields in arriving at the Faraday cages with different electric potentials and in departing from the ends of the Faraday cages to impact the distant phosphorescent screen. However, these end effects are independent of the transit time through the two Faraday cages with the given potential difference. Note, too, that the Faraday cages may, in principle, be arbitrarily long, providing an arbitrarily large phase shift, relative to which the end effects are negligible.

To obtain some idea of the magnitude of the Aharonov-Bohm effect, note that the phase shift is

$$\frac{2\pi S}{h} = 1.5 \times 10^7 \, \Phi \text{ radians}$$

where Φ is in maxwells. The shift amounts to $S/h = 2.5 \times 10^6 \, \Phi$ fringes, from which it is evident that the effect is detected by working with very small magnetic fluxes, so that the number of shifted fringes can be followed. The effect has been detected and verified in the laboratory, thanks to the ingenuity of several experimenters (cf. Rohrlich and Chambers 1960; Olariu and Popescu 1985; Tonomura et al. 1986) who developed tiny magnetic systems to project interference patterns, while the authors of the third reference enclosed the microsystem in a microsuperconducting shield to satisfy the critics that there were no stray magnetic fields.

4 Field Lines

4.1 Basic Considerations

Approximately a century and a half ago Michael Faraday pointed out the importance of the concept of field lines in visualizing the form of a field, electric or magnetic. We could add gravitational fields as well, although gravitational fields are generally only very closely radial in form and are not manipulated by the theorist beyond the static Newtonian gravitational state. The concept of field lines, carried bodily in the large-scale swirling plasma, is fundamental to the magnetohydrodynamics of the cosmos. The field lines of any vector field B_i make up the family of solutions of the two differential equations

$$\frac{dx}{B_x} = \frac{dy}{B_y} = \frac{dz}{B_z} \tag{4.1}$$

In terms of the arc length *ds* along a field line this can be written

$$\frac{dx_i}{ds} = \frac{B_i}{B} \tag{4.2}$$

where B represents that magnitude of the field B_i. The same equations with B_i replaced by E_i provide the field lines of an electric field, of course. The reader is presumably familiar with the radial field lines of a point charge, the closed lobe of field lines of a dipole, and the double lobes of field lines of a quadrupole, etc.

One may describe a field line as the path traced out by a point charge moving always in the direction of the local field B_i. Thus, a point charge, electric or magnetic, immersed in a viscous fluid, so that it is driven only very slowly by the force exerted on it by the field (the inertia of the moving particle and the entrained fluid being negligible), follows a field line. A small electric or magnetic dipole, allowed to swing freely about its position in space, points along the field at its location. Thus, a field line is mapped out if the dipole moves continuously in the direction in which it is pointing.

Now, as already noted, the scarcity of magnetic monopoles implies that $\nabla \cdot \mathbf{B} = 0$, as distinct from the condition $\nabla \cdot \mathbf{E} = 4\pi\delta$ for the electric field. Since field lines begin and end only at a charge, it follows that the magnetic field lines have no ends. In a two-dimensional space the magnetic field lines either close on themselves or extend to infinity. In the three-dimensional space of the cosmos field lines close on themselves only in fields with special symmetries that effectively reduce the field topology to two dimensions. So in almost all fields in nature the lines either circle endlessly, approaching an ergodic topology, or they extend to infinity. The likelihood that a field line close on itself is the vanishing probability that upon circling around one or more times the line hits itself exactly in the tail. The cross section for such an encounter is zero, of course (Parker 1979, pp. 275–279; 1994, pp. 101–104).

It is not uncommon to read or hear it said that $\nabla \cdot \mathbf{B} = 0$ implies that field lines are closed on themselves in three-dimensional space. The idea is durable, probably as a carryover from the fact that textbooks on magnetism are prone to discuss, and to draw, the lines for fields only of precise analytical form.

Having repeatedly remarked on the absence of magnetic monopoles, we note that for some purposes the open end of a long, thin, finely wound solenoid provides a useful local working approximation to a magnetic monopole. Denote the length of the solenoid by L and the small radius of the solenoid by ε. The essential point is that the magnetic field spreads out radially in all directions from the open end of the solenoid, sketched in Fig. 4.1. The field from the opposite end of the solenoid is small $O(\varepsilon^2/L^2)$ compared to the field in the open end, so the deviation from radial

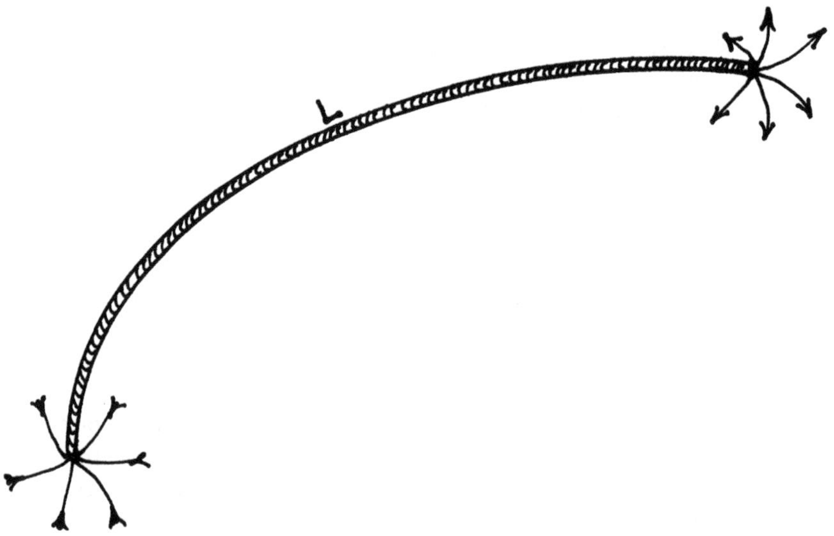

Fig. 4.1 A sketch of the radial magnetic field lines emanating from the ends of a long, thin, closely wound solenoid of length L large compared to its radius ε.

becomes significant only at a radial distance r from the open end that is a significant fraction of L. Over the range $\varepsilon < r \ll L$ the radial field intensity $B(r)$ is well approximated by $\Phi/4\pi r^2$, where Φ is the total magnetic flux contained within the solenoid and issuing from the open end.

It goes without saying that the family of solutions of eqn. (4.1) or (4.2) is everywhere dense, for the magnetic field is a continuum. This obvious point is not appreciated as universally as one might think. For instance, some years ago this author was invited to write an article on magnetic field lines, the topology of field lines in three-dimensional space, and the fundamental conceptual role they play in moving with the conducting fluid in MHD. It was suggested that the *American Journal of Physics* would be the appropriate venue for publication. The paper was written and submitted, and after a time it was returned by the editor with a firm rejection, based on the report of the referee, who was "an expert on the subject of magnetism." The referee remarked simply that the idea of field lines was ridiculous because anyone who knew anything about magnetism was aware that a magnetic field is a continuum and not made up of strings.

4.2 The Optical Analogy

The manner in which an individual field line extends through a region of field $B_i(\mathbf{r})$ is described by the *optical analogy* (Parker 1991, 1994, chapter 7) for the case that the field is curl free, i.e., when the field can be described as the gradient of a scalar. This applies to all electrostatic fields and to all regions of magnetic field that are curl free, i.e., free of electric current flowing perpendicular to the field. As we shall see in chapter 6, it applies in any flux surface of a so called *force-free* equilibrium magnetic field, in which the divergence $\partial M_{ij}/\partial x_j$ of the stress tensor M_{ij} is equal to zero, because the curl is parallel, rather than perpendicular, to the magnetic field.

Using a magnetic field B_i as an example, suppose that $B_i = -\partial\phi/\partial x_i$. Then the equations for a field line are

$$B\frac{dx_i}{ds} = -\frac{\partial\phi}{\partial x_i} \tag{4.3}$$

with

$$B^2 = (\nabla\phi)^2$$

This equation has the same form as the eikonal equation

$$n\frac{dx_i}{ds} = \frac{\partial\psi}{\partial x_i} \tag{4.4}$$

for the optical ray paths of the wave exp $i\psi$ in a medium with nonuniform index of refraction n (cf. Born and Wolf 1975). The eikonal equation follows from the fact that the ray path is in the direction of the gradient of the wave phase ψ, and the rate at which the phase of the wave progresses along the ray path is proportional to the wave number, and hence proportional to n. That is to say, $(\nabla\psi)^2 = n^2$.

Comparing eqns. (4.3) and (4.4) it is evident that a field line extends through the region of field as if it were an optical ray in an index of refraction proportional to the magnitude $B(\mathbf{r})$ of the field $B_i(\mathbf{r})$. Thus, the Snell refraction laws apply to the field line and Fermat's principle is applicable, stating that the field line connecting points 1 and 2 follows the path for which the integral of B is an extremum—usually a minimum:

$$\delta \int_1^2 ds B(\mathbf{r}) = 0 \tag{4.5}$$

It follows then that the Euler equations describe the path. The two-dimensional field $B\ (x, y)$ confined to a plane requires that

$$\delta \int_1^2 dx \left(1 + y'^2\right)^{1/2} B(x, y) = 0 \tag{4.6}$$

upon using x as the independent variable and with $y' \equiv dy/dx$. The Euler equation is

$$\frac{d}{dx}\left[\frac{y' B(x, y)}{(1 + y'^2)^{1/2}}\right] - (1 + y'^2)^{1/2}\frac{\partial B}{\partial y} = 0 \tag{4.7}$$

which can be written

$$\frac{y''}{1 + y'^2} + y'\frac{\partial \ln B}{\partial x} - \frac{\partial \ln B}{\partial y} = 0 \tag{4.8}$$

In polar coordinates (ϖ, φ) with ϖ as the independent variable,

$$\delta \int d\varpi (1 + \varpi^2 \varphi'^2)^{1/2} B(\varpi, \varphi) = 0 \tag{4.9}$$

with $\varphi' \equiv d\varphi/d\varpi$. The Euler equation becomes

$$\frac{d}{d\varpi}\left[\frac{\varpi^2 \varphi' B}{(1 + \varpi^2 \varphi'^2)^{1/2}}\right] - (1 + \varpi^2 \varphi'^2)\frac{\partial B}{\partial \varphi} = 0 \tag{4.10}$$

If, on the other hand, φ is the independent variable, then, with $\varpi' \equiv d\varpi/d\varphi$, we have

$$\frac{d}{d\varphi}\left[\frac{\varpi' B}{(\varpi^2 + \varpi'^2)^{1/2}}\right] - (\varpi^2 + \varpi'^2)^{1/2}\frac{\partial B}{\partial \varpi} = 0 \qquad (4.11)$$

Now when the field $B_i(\mathbf{r})$ is known, it is easiest to compute the field lines directly from eqn. (4.1) or (4.2). However, in MHD where the topology of the field, but not the precise form of $B_i(\mathbf{r})$, is known, the optical analogy can be an effective tool to treat the breaks in the flux surfaces and the associated surfaces of tangential discontinuity. It is also useful for treating the path of an individual slender flux tube that has been displaced by the motion of its footpoints (as at the surface of the Sun) from its initial path in the equilibrium ambient field, sketched in Fig. 4.2. For, assuming that the flux bundle has not been twisted, the

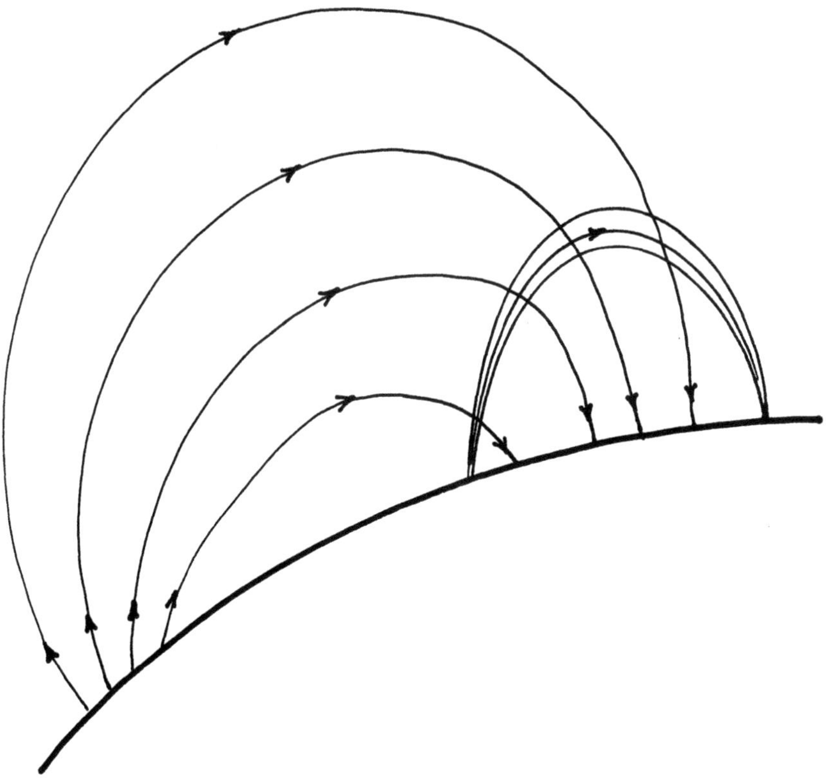

Fig. 4.2 A schematic drawing of a flux bundle displaced relative to the pattern of the ambient bipolar magnetic field.

internal field has no curl and is subject to the optical analogy. If the internal field is described by the scalar potential ϕ, then pressure balance between the internal field of the flux bundle and the external field B decrees that $(\nabla\phi)^2 = B^2$. If there is also a gas pressure P in the external field (in excess of the gas pressure in the flux bundle) that must be included as well, so that equilibrium requires $(\nabla\phi)^2 = B^2+8\pi P$. This is, of course, the eikonal equation for an equivalent index of refraction $(B^2 + 8\pi P)^{1/2}$ (cf. Parker 1994, pp. 219–224; 1996a, section 6.2). Note that if the displaced flux bundle has an internal gas pressure in excess of the external gas pressure, then the effective P may be negative. The concept of displaced flux bundles plays an important role in understanding the small-scale internal structure of the bipolar magnetic fields extending outward from the surface of the Sun, for it must be remembered that the photospheric footpoints of these large-scale fields at the visible surface consist of hundreds and thousands of separate tiny intense individual flux bundles carried about more or less independently of each other in the subsurface convection.

5 Maxwell's Equations

Having developed the properties of quasi-static (nonradiative) electric and magnetic fields, the next step is to assemble Maxwell's time-dependent equations (1.5) and (1.6). We have not hesitated to discuss Maxwell's equations in the development thus far, or, indeed, to employ Faraday's induction equation in discussing the quasi-static field equations. But it is only proper that we give detailed attention to the basis for Maxwell's final generalization of the electromagnetic equations.

Faraday's induction equation originated from his experimental discovery that the electromotive force V induced in a loop of wire is proportional to the time rate of change $d\Phi/dt$ of the total magnetic flux Φ through the loop. That is to say,

$$V = \frac{1}{c}\frac{d\Phi}{dt} \tag{5.1}$$

where we have written in the proportionality constant $1/c$ appropriate for cgs units. The emf V around a closed contour Γ is defined as

$$V = \oint_\Gamma d\mathbf{s}\cdot\mathbf{E}$$

where the integration is around the close contour Λ. Applying Stokes's theorem it follows that

$$V = \int_S d\mathbf{S}\cdot\nabla\times\mathbf{E}$$

where the integration is over the surface S enclosed by Γ. The time rate of change of the magnetic flux Φ through the surface S is

$$\frac{d\Phi}{dt} = \frac{d}{dt}\int_S d\mathbf{S}\cdot\mathbf{B}$$

For a fixed loop Γ, d/dt can be replaced by $\partial/\partial t$, and Faraday's experimental result becomes

$$\int_S d\mathbf{S}\cdot\nabla\times\mathbf{E} = -\frac{1}{c}\int_S d\mathbf{S}\cdot\frac{\partial\mathbf{B}}{\partial t} \tag{5.2}$$

where the minus sign is introduced based on the experimental direction of the emf. This integral relation is valid for all closed contours Γ, which is possible only if the integrands are equal at every point. Hence, we require that

$$\frac{\partial \mathbf{B}}{\partial t} = -c\nabla \times \mathbf{E} \tag{5.3}$$

at every point. This is eqn. (1.5), of course. It is the differential form of Faraday's law of induction. The integral form is given by eqn. (5.1), representing the direct experimental measurement.

Now to deduce Maxwell's equation (1.6), note that conservation of electric charge δ in a time-varying electric field requires that

$$\nabla \cdot \mathbf{j} = -\frac{\partial \delta}{\partial t} \tag{5.4}$$

instead of $\nabla \cdot \mathbf{j} = 0$ for steady conditions. Thus, eqn. (3.10) becomes

$$[\mathbf{j}(\mathbf{r}') \cdot \nabla] \frac{\mathbf{r} - \mathbf{r}'}{|\mathbf{r} - \mathbf{r}'|^3} = -\frac{\partial}{\partial x'_k}\left[j_k(\mathbf{r}') \frac{\mathbf{r} - \mathbf{r}'}{|\mathbf{r} - \mathbf{r}'|^3} \right] + \frac{\mathbf{r} - \mathbf{r}'}{|\mathbf{r} - \mathbf{r}'|^3} \frac{\partial \delta(\mathbf{r}')}{\partial t} \tag{5.5}$$

Note again that the first term on the right-hand side represents a divergence, which vanishes upon integration over the entire volume. So upon integration over the whole $\mathbf{r}'$ space, there is only the contribution

$$\begin{aligned} \frac{1}{c} \int d^3\mathbf{r}' \frac{\mathbf{r} - \mathbf{r}'}{|\mathbf{r} - \mathbf{r}'|^3} \frac{\partial \delta(\mathbf{r}')}{\partial t} &= \frac{1}{c} \frac{\partial}{\partial t} \int d^3\mathbf{r}' \delta(\mathbf{r}', t) \frac{\mathbf{r} - \mathbf{r}'}{|\mathbf{r} - \mathbf{r}'|^3} \\ &= +\frac{1}{c} \frac{\partial \mathbf{E}(\mathbf{r}, t)}{\partial t} \end{aligned} \tag{5.6}$$

upon using eqn. (2.4). This extra term must be added to eqn. (3.11), so that it becomes

$$c\nabla \times \mathbf{B}(\mathbf{r}, t) = 4\pi \mathbf{j}(\mathbf{r}, t) + \frac{\partial \mathbf{E}(\mathbf{r}, t)}{\partial t} \tag{5.7}$$

recognizable as Maxwell's equation (1.6). Thus, Maxwell's equation is implied by the Biot-Savart integral given that charge is conserved in a time-varying field.

Maxwell arrived at the equation by a different route, introducing the displacement current $(1/4\pi)\partial\mathbf{E}/\partial t$ as a hypothesis. He thought of it as representing the electric current associated with the time rate of change of the polarization of the ether, in direct analogy to the displacement of charge, i.e., the electric current, $\partial\mathbf{\Pi}/\partial t$ associated with the time rate of change of the dipole moment $\mathbf{\Pi}$ per unit volume in a material dielectric. From that point of view, Maxwell thought of his equation as the generalization of Ampere's law, eqn. (3.5), to include the entire current $\mathbf{L}(\mathbf{r}, t) = \mathbf{j}(\mathbf{r}, t) + (1/4\pi)\partial\mathbf{E}/\partial t$. The addition of the displacement current implies charge conservation, of course, as may be seen from the divergence of eqn. (1.6) or (5.7). We find here that the elementary concept of charge conservation, described by eqn. (5.4), automatically inserts the displacement current into the Biot-Savart integral form of Ampere's law to give the complete Maxwell equation.

It is curious that this connection is not generally pointed out in the standard textbooks. It is not entirely unknown, being rediscovered and discussed from time to time (cf. Mello 1972; Biswas 1988). It leads to interesting questions on current closure and whether the displacement current should be included in the Biot-Savart integral. The fact is that if all moving electric charges are recognized, including the elastically bound charges associated with the polarization of a dielectric, then the Biot-Savart integral provides the magnetic field in terms of the total current density $\mathbf{j}(\mathbf{r}, t)$. On the other hand, if we apply Stokes theorem to Maxwell's equation, integrating over an area S defined by the closed contour Γ, we obtain

$$c\oint_{\Gamma} d\mathbf{S}\cdot\mathbf{B} = 4\pi\int_{S} d\mathbf{S}\cdot\mathbf{j} + \int_{S} d\mathbf{S}\cdot\frac{\partial\mathbf{E}}{\partial t}$$

and the displacement current is included and cannot be neglected in many cases (see discussion by French and Tessman 1963; Bartlett 1990; Vasyliunas 1999). The essential point is that the two approaches to calculating the magnetic field are exactly equivalent, the one derived from the other.

Let us go back to the Biot-Savart integral, then, and ask how it is that this integral relation between static $\mathbf{j}$ and $\mathbf{B}$ introduces the so called displacement current $(1/4\pi)\partial\mathbf{E}/\partial t$, which is, after all, the essential ingredient for electromagnetic radiation in a vacuum. The first point to recognize is that the current $\mathbf{j}$, and hence the Biot-Savart integral, vanish in the absence of electric charges, so that the integral implies nothing for the vacuum. The integral has physical significance only in the presence of moveable electric charges, i.e., matter, such as a dielectric or a plasma, or an electric current driven in a wire by an applied emf. Then $\mathbf{j}(\mathbf{r}, t)$ is not identically zero. Second, we assume that the static relation represented by the Biot-Savart integral is valid for a field and current varying slowly with time. With the concept of charge conservation this assumption requires that $\nabla \cdot \mathbf{j} = -\partial\delta/\partial t$, thereby introducing the small term $\partial\delta/\partial t$, and ultimately the small correction $\partial \mathrm{E}/\partial t$ into the differential form of Ampere's law. Thus, with the addition of the idea of charge conservation in slowly varying fields, the Biot-Savart integral thrusts Maxwell's hypothesis of the displacement current upon us. The interesting thing is, then, that nothing further needs to be added to the electromagnetic equations when we turn to rapidly varying fields and radiative processes and propagation. The rapidly varying fields satisfy exactly the same equations as the slowly varying fields.

The fact of the equivalence of the Biot-Savart integral and Maxwell's equation provides some amusement when we look back at the history of the development of Maxwell's equation. There was a general disbelief of Maxwell's equation in England following its introduction by Maxwell in 1865. For instance, Kelvin asserted that he did not understand the equation, which we may accept as quite likely. He went on to declare that he did not need Maxwell's equation because light could just as well be explained by an elastic ether, so there was no benefit to be derived from Maxwell's equation. Today we would note the welcome simplification, that, given the displacement current, the electric and magnetic fields can stand on their own to provide light, and the burdensome concept of the ether can be discarded as superfluous. Now in spite of Kelvin's disclaimer of Maxwell, we may guess that Kelvin and the other disbelievers accepted the Biot-Savart integral, without recognizing that it implies Maxwell's equation. Denial is always a risky business. You never know when you may be standing on the same rug that you propose to yank out from under someone else.

On the other hand, it should be emphasized that Oliver Heaviside recognized the profound significance of Maxwell's equation and, indeed, contributed to pruning away some of the excess baggage attached to Maxwell's original concepts. On the Continent, Helmholtz and Rowland grasped the importance. Between 1885 and 1889 Hertz, inspired by the implications of Maxwell's equation, carried out his celebrated experiments, producing, detecting, and measuring the velocity of propagation of radio waves in the laboratory, and identifying radio waves as the same phenomenon as light waves, all with speed c equal to the ratio of the cgs emu unit of charge to the cgs esu unit (employed in the present *Conversations*). Yet the 1894 edition of the *Encyclopedia Britannica* has an extensive article on electricity and another on magnetism, which mention electrodynamics and Maxwell's earlier contributions but give no hint of Maxwell's equation and its implications for electromagnetic waves. Only the author of the biographical article on Maxwell remarks on Maxwell's fundamental contribution, noting that the primary test of the electromagnetic theory of light depends on whether the velocity of light is accurately given by the ratio of the units of charge in emu and esu. That author enthusiastically refers the reader to Maxwell's *Treatise on Electricity and Magnetism* (1873).

So the history of the people in science is always fascinating, and often amusing when viewed after another century of scientific advance. Each century gets a few laughs at some of the antics of the preceding century, and is itself a subject of humor for the following century.

6 Maxwell and Poynting

6.1 Poynting's Momentum and Energy Theorems

We owe to Poynting the formal demonstration of the mutual consistency of Newton's momentum equation with Maxwell's equations for electromagnetic fields. Indeed, given that both sets of equations are fundamental laws of nature, they must be mutually consistent. Or, to put it differently, the physical concepts associated with one must be brought into line with the physical concepts associated with the other. Maxwell's equations are correct in the relativistic regime, while Newton's equations are not. So their compatibility in the present context is treated only in the nonrelativistic regime ($v << c$) where both are valid.

Suppose, then, that the mass density is described by the continuous function $\rho(\mathbf{r}, t)$. At the fluid level ρ would be a smooth function, varying for our purposes only over some macroscopic scale. On the other hand, if we view the plasma, or other matter, on a microscopic scale, then ρ is a collection of classical fundamental particles, in the form of many, many moving spheres, with characteristic diameters of the order of 10^{-13} cm or more. Each particle is a spike in ρ in three-dimensional space with the density ρ between particles equal to zero. Similarly, the charge density $\delta(\mathbf{r}, t)$is introduced with spikes at the position of each particle. The spikes in the mass density $\rho(\mathbf{r}, t)$ and spikes in the charge density $\delta(\mathbf{r}, t)$ are tied to each other by forces other than electromagnetic, i.e., within a particle both mass density and charge density are somehow held together in opposition to the strong electric and magnetic forces tending to disperse $\delta(\mathbf{r}, t)$. So the particle structure is fixed, with $\delta(\mathbf{r}, t)$ and $\rho(\mathbf{r}, t)$ spiky but everywhere coincident, continuous and bounded.

In terms of ρ and δ the nonrelativistic momentum equation is

$$\rho(\mathbf{r}, t)\frac{d\mathbf{v}}{dt} = \delta(\mathbf{r}, t)\left[\mathbf{E}(\mathbf{r}, t) + \frac{\mathbf{v} \times \mathbf{B}(\mathbf{r}, t)}{c}\right] \tag{6.1}$$

where $\mathbf{E} + \mathbf{v} \times \mathbf{B}/c$ is the electric field experienced by the particle, as described in the succeeding chapter. In this equation the particle velocity $\mathbf{v}$ refers to the velocity of the spike in ρ and δ representing the particle.

The energy equation is obtained from the scalar product of **v** with the momentum equation, yielding

$$\tfrac{1}{2}\rho(\mathbf{r}, t)\frac{dv^2}{dt} = \delta(\mathbf{r}, t)\mathbf{v}\cdot[\mathbf{E}(\mathbf{r}, t) + \mathbf{v} \times \mathbf{B}(\mathbf{r}, t)] \tag{6.2}$$

Poynting's theorems, defining the electromagnetic energy density, energy flux, and momentum density, follow from the standard textbook manipulations of these two expressions, using the familiar vector identities.

Note that the electric current density is $\mathbf{j}(\mathbf{r}, t) = \delta(\mathbf{r}, t)\mathbf{v}$, while the term $\mathbf{v} \cdot (\mathbf{v} \times \mathbf{B})$ is identically zero, of course. With the aid of eqn. (5.7) and then eqn. (5.3), eqn. (6.2) can be rewritten

$$\begin{aligned}
\tfrac{1}{2}\rho\frac{dv^2}{dt} &= \mathbf{j}\cdot\mathbf{E} \\
&= \frac{c}{4\pi}\mathbf{E}\cdot(\nabla \times \mathbf{B}) - \frac{\partial}{\partial t}\frac{E^2}{8\pi} \\
&= \frac{c}{4\pi}\mathbf{B}\cdot\nabla \times \mathbf{E} - \frac{c}{4\pi}\nabla\cdot(\mathbf{E} \times \mathbf{B}) - \frac{\partial}{\partial t}\frac{E^2}{8\pi} \\
&= -\frac{1}{4\pi}\mathbf{B}\cdot\frac{\partial \mathbf{B}}{\partial t} - \nabla\cdot\left[\frac{c}{4\pi}(\mathbf{E} \times \mathbf{B})\right] - \frac{\partial}{\partial t}\frac{E^2}{8\pi} \\
&= -\frac{\partial}{\partial t}\frac{E^2 + B^2}{8\pi} - \nabla\cdot\mathbf{P}
\end{aligned} \tag{6.3}$$

where the Poynting vector **P** is

$$\mathbf{P} = c\,\frac{\mathbf{E} \times \mathbf{B}}{4\pi} \tag{6.4}$$

The energy eqn. (6.3) can then be rewritten as

$$\tfrac{1}{2}\rho\frac{\partial v^2}{\partial t} + \tfrac{1}{2}\rho v_j\frac{\partial v^2}{\partial x_j} + \frac{\partial}{\partial t}\frac{E^2 + B^2}{8\pi} + \frac{\partial P_j}{\partial x_j} = 0 \tag{6.5}$$

Then multiply the equation

$$\frac{\partial \rho}{\partial t} + \frac{\partial \rho v_j}{\partial x_j} = 0$$

for conservation of matter by $v^2/2$ and add it to the energy equation, obtaining

$$\frac{\partial U}{\partial t} = -\nabla \cdot \mathbf{P} - \nabla \cdot \left(\tfrac{1}{2}\rho v^2 \mathbf{v} \right) \tag{6.6}$$

where U is the energy density,

$$U = \tfrac{1}{2}\rho v^2 + \frac{E^2 + B^2}{8\pi} \tag{6.7}$$

Equation (6.6) states that the time rate of change of the energy density U is equal to the negative divergence of the total energy flux $\mathbf{P} + \frac{1}{2}\rho v^2 \mathbf{v}$. That is to say, it is evident from the formal structure of eqn. (6.6) that we must interpret the Poynting vector **P** as the energy flux in the electric and magnetic fields, with field energy densities $E^2/8\pi$ and, $B^2/8\pi$, respectively. We have already shown, from simple mechanical manipulations of electric and magnetic fields, why $E^2/8\pi$ and $B^2/8\pi$ must be thought of as the energy densities. The bottom line is that Newton and Maxwell are compatible and we have learned that the electromagnetic energy flux is given by the Poynting vector **P**.

At first sight this seems counterintuitive, because the Poynting vector tells us that the energy transport along a power line takes place in the space around the line, whereas we are in the habit of thinking of the energy being transported by the electric current within the line. We define an electric potential, viz. the voltage V applied to the line, and think of each conduction electron as having a potential energy $-eV$. That point of view handles the power transmission properly, but the "up close" physics turns out to be different. Instead of the vague notion of a potential energy $-eV$, we have the detailed picture that the energy is transported in the electric and magnetic fields associated with the wire.

This is perhaps the appropriate place to note that the foregoing equations are not altered when, in the nonrelativistic case, they are averaged over scales large compared to the interparticle distances, but small compared to the scale of variation of the macroscopic fields. The local Coulomb fields of the individual ions and electrons drop out, because their energy is generally considered part of the rest energy of the individual particles themselves. Thus, the large-scale fields are continuous and smooth across the individual particle spikes in ρ and δ.

The next step is to start with the momentum equation (6.1), noting that the charge density can be eliminated with the aid of eqn. (2.10). The result can be written

$$\rho \frac{dv}{dt} = \frac{\mathbf{E}\nabla \cdot \mathbf{E}}{4\pi} + \frac{\mathbf{B}\nabla \cdot \mathbf{B}}{4\pi} + \frac{\mathbf{j} \times \mathbf{B}}{c} \tag{6.8}$$

upon adding the term in $\nabla \cdot \mathbf{B}$ (equal to zero) to preserve symmetry. Then, using eqn. (5.7) to eliminate $\mathbf{j}$, it follows that

$$
\begin{aligned}
\rho \frac{dv}{dt} &= \frac{\mathbf{E}\nabla\cdot\mathbf{E} + \mathbf{B}\nabla\cdot\mathbf{B}}{4\pi} + \frac{(\nabla \times \mathbf{B}) \times \mathbf{B}}{4\pi} - \frac{\partial \mathbf{E}}{\partial t} \times \frac{\mathbf{B}}{4\pi c} \\
&= \frac{\mathbf{E}\nabla\cdot\mathbf{E} + \mathbf{B}\nabla\cdot\mathbf{B}}{4\pi} + \frac{(\nabla \times \mathbf{B}) \times \mathbf{B}}{4\pi} - \frac{\partial}{\partial t}\frac{\mathbf{E} \times \mathbf{B}}{4\pi c} \\
&\quad - \frac{\partial \mathbf{B}}{\partial t} \times \frac{\mathbf{E}}{4\pi c} \\
&= \frac{\mathbf{E}\nabla\cdot\mathbf{E} + \mathbf{B}\nabla\cdot\mathbf{B}}{4\pi} + \frac{(\nabla \times \mathbf{B}) \times \mathbf{B} + (\nabla \times \mathbf{E}) \times \mathbf{E}}{4\pi} \\
&\quad - \frac{\partial}{\partial t}\frac{\mathbf{P}}{c^2}
\end{aligned}
\tag{6.9}
$$

upon using eqn. (5.3). Then with the vector identity

$$(\nabla \times \mathbf{A}) \times \mathbf{A} = (\mathbf{A}\cdot\nabla)\mathbf{A} - \nabla\tfrac{1}{2}A^2$$

this can be written

$$
\begin{aligned}
\rho \frac{dv}{dt} + \frac{\partial}{\partial t}\frac{\mathbf{P}}{c^2} &= \frac{\mathbf{E}\nabla\cdot\mathbf{E} + (\mathbf{E}\cdot\nabla)\mathbf{E} + \mathbf{B}\nabla\cdot\mathbf{B} + (\mathbf{B}\cdot\nabla)\mathbf{B}}{4\pi} \\
&\quad - \nabla\left(\frac{E^2 + B^2}{8\pi}\right)
\end{aligned}
\tag{6.10}
$$

In terms of the stress tensor

$$M_{ij} = -\delta_{ij}\frac{E^2 + B^2}{8\pi} + \frac{E_iE_j + B_iB_j}{4\pi} \tag{6.11}$$

we have

$$\rho \frac{dv_i}{dt} + \frac{\partial}{\partial t}\frac{P_i}{c^2} = \frac{\partial M_{ij}}{\partial x_j} \tag{6.12}$$

The stress tensor is obviously the Maxwell stress tensor that we have already obtained from simple physical considerations in eqns. (2.15), (3.1), and (3.22).

The left-hand side of eqn. (6.12) can be rewritten by multiplying the continuity equation by v_i and adding to eqn. (6.12), obtaining

$$\frac{\partial}{\partial t}\left(\rho v_i + \frac{P_i}{c^2}\right) = \frac{\partial M_{ij}}{\partial x_j} - \frac{\partial}{\partial x_j}\rho v_i v_j \tag{6.13}$$

It is evident that $\rho v_i + P_i/c^2$ represents the total momentum density with P_i/c^2 the momentum density carried by the electromagnetic field. Note then that the momentum flux density is $M_{ij} - \rho v_i v_j$, the first term representing the Maxwell stress and the second term representing the flux of momentum density transported by the bulk motion v_i of the particles or fluid. Note that the electromagnetic energy flux P_i is c^2 times the momentum density in the electromagnetic field. That is to say, for a given energy flux, the electromagnetic momentum density is small $O(v^2/c^2)$ and can be neglected in nonrelativistic dynamics.

6.2 Applications

In magnetohydrodynamics, where $\mathbf{E} = -\mathbf{v} \times \mathbf{B}/c$, the electromagnetic energy flux $\mathbf{P}$ turns out to be $\mathbf{v}_\perp B^2/4\pi$, where $\mathbf{v}_\perp$ is the bulk plasma velocity perpendicular to the magnetic field, and $B^2/4\pi$ represents the magnetic enthalpy density. The electromagnetic momentum density is $v_\perp(B^2/4\pi c^2)$, to be compared with the momentum density $\rho\mathbf{v}$ carried by the plasma. It is evident that only if the field were so strong that the Alfvèn speed, $B/(4\pi\rho)^{1/2}$, for the propagation of large-scale electromagnetic waves in a plasma, were comparable to c would the electromagnetic momentum be significant, i.e., only if $B^2/8\pi \approx \rho c^2/2$, so that the energy density of the field is comparable to the rest energy of the particles or plasma.

An extreme example would be a magnetar (magnetic neutron star) with a magnetic field estimated at $B = 10^{15}$G, in order of magnitude, for which the energy density is comparable to the rest energy density of $\rho \approx 4 \times 10^7$g/cm^3. Within the magnetar the characteristic density is comparable to nuclear densities of the general order of 10^{14}g/cm^3, so that $B^2/8\pi$ is actually small compared to ρc^2. However, the characteristic scale of the external magnetic field of the neutron star is comparable to the radius of the star, whereas the density ρ cuts off rapidly at the surface and ρc^2 quickly falls below $B^2/8\pi$ so that exterior MHD becomes relativistic and the displacement current $\partial\mathbf{E}/\partial t$ cannot be neglected. The dynamics becomes much more complex than in the familiar nonrelativistic MHD of the rest of the world.

6.3 Electric and Magnetic Fields in Matter

In the presence of dielectric substances it is sometimes convenient to introduce the *electric displacement vector* **D**, defined as

$$\mathbf{D} = \mathbf{E} + 4\pi\Pi \tag{6.14}$$

where $\mathbf{\Pi}$ is the dipole moment per unit volume arising from the electrical polarization of the dielectric material by the electric field $\mathbf{E}$. Each molecule is deformed by $\mathbf{E}$ into a tiny electric dipole, say with a dipole moment m. If there are N such polarized molecules per unit volume, then $\mathbf{\Pi} = Nm$. A nonuniform polarization produces a local electric charge density $\delta = \nabla \cdot \mathbf{\Pi}/4\pi$ because the "heads" and "tails" of the microscopic dipoles are not entirely canceled by the opposite charges of the tails and heads, respectively, of the microscopic dipoles in front and behind. Hence, an electric field $\mathbf{E}$ imposed perpendicular to the face of a slab of dielectric results in surface charge densities $+\Pi$ and $-\Pi$ on opposite faces of the slab. Thus, the electric field $\mathbf{E}_D$ within the dielectric is reduced by the amount $\Delta\mathbf{E} = 4\pi\mathbf{\Pi}$ that terminates at the surface charge $\mathbf{\Pi}$, sketched in Fig. 6.1. Thus $\mathbf{D} = \mathbf{E}$ in the free space outside the dielectric and $\mathbf{D} = (\mathbf{E} - \Delta\mathbf{E}) + 4\pi\mathbf{\Pi} = \mathbf{E}$ inside the slab of dielectric. So $\mathbf{D}$ has the convenient property of being continuous across the boundary of the dielectric. The electric displacement $\mathbf{D}$ responds only to free charges δ_f, the bound charges being recognized by the inclusion of $\mathbf{\Pi}$, with

$$\nabla \cdot \mathbf{D} = 4\pi\delta_f \tag{6.15}$$

Turning to magnetic material, note that applying an external magnetic field $\mathbf{B}$ aligns the individual circulating currents of the electron spin, orbiting electrons, and molecular magnetic moments to provide

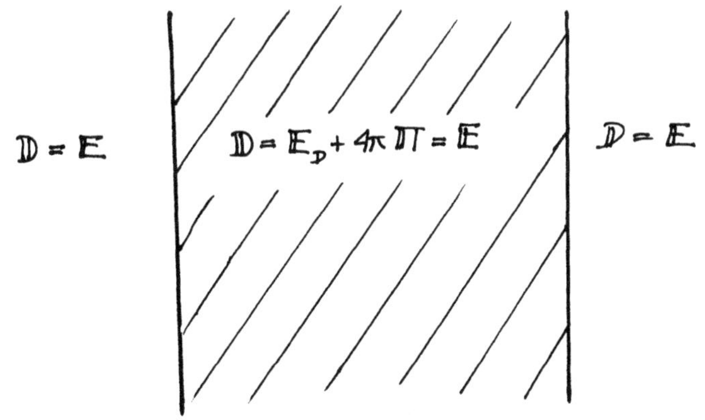

Fig. 6.1 A schematic drawing of the field $\mathbf{D} = \mathbf{E}_D + 4\pi\mathbf{\Pi}$ in a slab of dielectric as a consequence of the externally applied field $\mathbf{D} = \mathbf{E}$. The field D is the same both inside and outside the dielectric.

the net circulating current. Each current I circulating around a microscopic area A provides a dipole moment IA/c in the direction perpendicular to the surface A, indicated by the unit vector **e**. Summing over a unit of volume containing N such dipoles provides a net magnetic moment $\mathbf{M} = \mathbf{e}NIA/c$ per unit volume. This local magnetization, or current circulation, has a magnetic flux $4\pi\mathbf{M}$ associated with it. Thus, if in some way we could freeze the local **M** in place and switch off the externally applied magnetic field, there would remain the magnetic field of $4\pi\mathbf{M}$ of the local current circulation. So it is sometimes useful to write the total field **B** as the sum of $4\pi\mathbf{M}$ plus the vector **H**, where **H** is that part of the total **B** associated with electric currents circulating elsewhere. Thus, one writes

$$\mathbf{B} = \mathbf{H} + 4\pi\mathbf{M} \tag{6.16}$$

Note that the motivation for the decomposition of the magnetic field is different from that for decomposition of the electric field. For the electric field we added $4\pi\mathbf{\Pi}$ to the electric field **E** to obtain the field **D** that had a desirable property. In the magnetic case we simply decomposed the field **B** into two parts, based on the location of the electric currents with which the magnetic field **B** is associated.

We note in passing that the energy density associated with an electric field **E** is $\mathbf{E} \cdot \mathbf{D}/8\pi$, which includes the energy of the polarized molecules of the dielectric. The energy in the electric field alone is, of course, $E^2/8\pi$. The energy density of the polarization can be written $\mathbf{\Pi} \cdot \mathbf{E}/2$.

For a magnetic material the energy density is $\mathbf{B} \cdot \mathbf{H}/8\pi$ with the energy density $B^2/8\pi$ in the magnetic field itself, and $-\mathbf{M} \cdot \mathbf{B}/2$ in the orientation of the dipole moment per unit volume.

The magnetic vector **H** is called the *magnetic intensity* and **B** is called the *magnetic density*. The motivation for this terminology seems to be that the application of a magnetic field **H** to a volume of magnetic material polarizes the material medium, causing the additional magnetic field $4\pi\mathbf{M}$ to arise within the material. Thus, **H** is looked upon as the driving magnetic force, with the net result **B**. There is sometimes a discussion of which field, **B** or **H**, is the fundamental magnetic field, overlooking the fact that **B** is the magnetic field, while **H** is only part of it when magnetic material is present. In the absence of matter $\mathbf{B} = \mathbf{H}$ and there is no distinction between the two.

The essential point for our conversation of electric and magnetic fields in the cosmos is that the hot gases that are everywhere in the cosmos have very little electric polarizability and very little magnetic susceptibility, so $\mathbf{D} = \mathbf{E}$ and $\mathbf{H} = \mathbf{B}$ to good approximation.

6.4 SI Units

Here we confront the complications introduced when the electromagnetic field is expressed in SI (mks) units. The motivation for the SI units is the inconveniently small unit of charge and the inconveniently large unit of electric potential in the cgs electrostatic (esu) system when it comes to expressing the currents and voltages in electric circuits. Thus, for instance, the radiation theory of an antenna uses esu, whereas the theory of the electric circuit that drives the antenna is formulated in terms of coulombs, amperes, volts, and ohms. The esu system is not at all convenient for the electric circuit. One ampere is equal to 3×10^9esu units of charge, while one volt is equal to 1/300 of the esu statvolt (cf. Stratton 1941, pp. 17, 18).

Now, in switching from cgs to mks units it was pointed out in chapter 2 that the natural unit of charge in the mks system would be defined as the charge repelling an equal charge with a force of one newton at a distance of one meter. That gives an mks unit of charge equal to $10^{9/2}$ esu, but that does not fit the coulomb at 3×10^9 esu, and it gets no better when we define the natural mks unit of electric field as the field at a distance of one meter from the mks unit of charge ($10^{9/2}$ esu), giving a field intensity unit equal to $10^{1/2}$ esu = 3.16 statvolts/cm, i.e., about 10^3 V/cm. The systems just do not mesh in any simple way.

So the SI units represent a whole new system, employing the mks units of meters, kilograms, and seconds, but also insisting that the coulomb is the basic unit of charge with no theoretical relation to mass, length, and time. Electromagnetic theory in SI units is formulated in diverse ways (compare Stratton 1941 and Shadowitz 1988), using the four vectors **E**, **D**, **H**, and **B** to represent the electromagnetic field in vacuum. Coulomb's law for the force between charges q_1 and q_2 becomes

$$F = \frac{q_1 q_2}{4\pi\varepsilon_0 r^2} \tag{6.17}$$

where ε_0 is a constant. With $q_1 = q_2 = 1$ C $= 3 \times 10^9$ esu, $r = 1$ m = 100 cm, and $F = 1$ N $= 10^5$ dynes, it follows that $4\pi\varepsilon_0 = 1/9 \times 10^9$ SI units. It is convenient then to define the vector $\mathbf{D} = \varepsilon_0\mathbf{E}$, with ε_0 called the *permitivity of the vacuum*. The vector **E** is called the *electric field intensity* and **D** is called the *electric displacement*. To fit in with these definitions and to keep Maxwell's equations in tidy form, the constant μ_0 is introduced in association with electric currents and magnetic fields, writing $\mathbf{B} = \mu_0\mathbf{H}$ with μ_0 referred to as the *permeability of the vacuum*. The vector **H** is again called the *magnetic field intensity* and **B** the *magnetic* *induction*. Then it turns out that μ_0 is subject to the constraint that $\mu_0 \varepsilon_0 = 1/c^2$. Since charge is not defined in terms of meters, kilograms, and seconds, the SI system is obliged to introduce the unit of charge Q as an independent unit in addition to the usual length L, mass M, and time T. It turns out, then, that the permittivity ε_0 has dimensions T^2Q^2/ML^3 and the permeability μ_0 has dimensions ML/Q^2. So **E** has dimensions ML/QT^2 and **D** weighs in with Q/L^2. The magnetic induction **B** has dimensions M/QT and the intensity **H** of that same field has dimensions Q/LT. With these definitions in mind it is not obvious how to explain the physical significance of **D** relative to **E** and **H** relative to **B**, as was easily done for the cgs system in section 6.3. Nor can any physical significance be assigned to ε_0 and μ_0. Indeed, one may ask which of **H** or **B** is the true magnetic field? Maxwell's equations for fields in a vacuum become

$$\frac{\partial \mathbf{B}}{\partial t} = -\nabla \times \mathbf{E} \qquad \nabla \cdot \mathbf{B} = 0$$

$$\mathbf{j} + \frac{\partial \mathbf{D}}{\partial t} = +\nabla \times \mathbf{H} \qquad \nabla \cdot \mathbf{D} = \delta$$

involving the four vectors **E**, **D**, **B**, and **H** to describe the two fields. Given the unclear physical significance of **D** relative to **E** and **H** relative to **B**, the physical connection between the electric and the magnetic fields is not obvious. The interpretation that "any change of either field with the passage of time is accompanied by a proportionate curl of the other, and vice versa," so conspicuous in eqns. (1.1) and (1.2), is not obvious here in SI country.

Perhaps the most obvious disadvantage of the SI system is the obscuration of the electric–magnetic symmetry of Maxwell's equations. How could the equations possibly be symmetric in electric and magnetic fields when electric and magnetic fields have different dimensions? Before going on to other difficulties, it is worthwhile exploring some of the implications of this absence of symmetry with respect to **E** and **B**. With different dimensions there is no convenient way to compare the strengths of electric and magnetic fields. Suppose that we have an electric field of one statvolt/centimeter and a magnetic field of one gauss. These two fields have the same stress and energy density. So their strengths are the same, and $E = B$ in cgs esu. On the other hand, in SI units the electric field is measured in volts/meter. With one statvolt equal to 300 V, the unit of electric field strength, volts/meter, is $1/(3 \times 10^4)$ of a statvolt/centimeter, so the unit electric field in esu has a value $E = 3 \times 10^4$ V/m in SI units. In contrast, the SI unit of magnetic field is the tesla, equal to 10^4 G, so the unit magnetic field of 1 G in esu has a value of 10^{-4} T in SI units. It follows that in

SI units $E/B = 3 \times 10^8$m/s, which we recognize as the speed of light c. So in SI units the electric and magnetic field units are way out of line with each other. The unit of magnetic field is immense compared to the unit of electric field.

This is more than an inconvenience. Stratton (1941, pp. 79, 80), in his enthusiastic presentation of electromagnetic theory in SI units, falls victim to this distortion in his discussion of Lorentz transformations. Again anticipating our own discussion of Lorentz transformations in chapter 6, we note from eqn. (7.1) that in cgs esu the nonrelativistic Lorentz transformations take the symmetric form

$$\mathbf{E}' = \mathbf{E} + \frac{\mathbf{v} \times \mathbf{B}}{c} \qquad \mathbf{B}' = \mathbf{B} - \frac{\mathbf{v} \times \mathbf{E}}{c} \tag{6.18}$$

where $\mathbf{E}'$ and $\mathbf{B}'$ represent the electric and magnetic fields, respectively, in the frame of reference moving with velocity $\mathbf{v}$ relative to the reference frame in which the fields are $\mathbf{E}$ and $\mathbf{B}$. In SI units the nonrelativistic Lorentz transformations become

$$\mathbf{E}' = \mathbf{E} + \mathbf{v} \times \mathbf{B} \qquad \mathbf{B}' = \mathbf{B} - \frac{\mathbf{v} \times \mathbf{E}}{c^2} \tag{6.19}$$

as Stratton's eqn. (111) of chapter 1. Stratton notes the c^2 in the denominator of the term $\mathbf{v} \times \mathbf{E}/c^2$ on the right-hand side. He then proclaims the term to be small $O(v^2/c^2)$ compared to the first term $\mathbf{B}$, so that the second term may be neglected, leaving the Lorentz transformation of the magnetic field as

$$\mathbf{B}' = \mathbf{B} \tag{6.20}$$

This is a curious conclusion, because in esu the term $\mathbf{v} \times \mathbf{E}/c$ is *not* negligible compared to $\mathbf{B}$. Stratton fell victim to the grotesque asymmetry of the SI units, forgetting that the terms in $\mathbf{E}$ and $\mathbf{B}$ cannot be compared directly because they have different dimensions. In esu $\mathbf{E}$ and $\mathbf{B}$ have the same dimensions, $M^{1/2}/LT$, while in SI units they have dimensions ML/QT^2 and M/QT, respectively, as already noted. Given that $\mathbf{E}$ in the numerator of $\mathbf{v} \times \mathbf{E}/c^2$ is inflated relative to $\mathbf{B}$ by the factor c, it is clear that $\mathbf{v} \times \mathbf{E}/c^2$ is small compared to $\mathbf{B}$ only to first order in v/c, not second order. So, even the experts on SI units are sometimes confounded by the obscurity of the system. We recall that the stated motive for introducing the SI units was to avoid the inconveniently small unit of charge in the cgs esu system when dealing with electric circuits. However, the penalty to be paid for going to coulombs as an independent entity is a heavy one.

Finally, consider again why the SI system is obliged to introduce the electric charge as a fourth dimensional quantity, in addition to M, L, and T. So far as mechanics and dynamics are concerned, it is apparent from Newton's equation of motion that the physics is covered by M, L, and T. Then when we move on to electric and magnetic fields, and electric charges, the properties of the fields and charges are conveniently defined in terms of their dynamical properties, for which M, L, and T are entirely adequate. The extra dimension Q in the SI system arises because of the insistence on the coulomb as the unit of charge, requiring the introduction of ε_0 into Coulomb's law, eqn. (6.17). That is to say, Coulomb's law relates charge to M, L, and T with the dimensions $M^{1/2}L^{3/2}/T$. If we choose to ignore this, requiring Q to be an independent unit on its own, then we have to undo the Coulomb law connection to $M^{1/2}L^{3/2}/T$ by introducing the additional factor ε_0. We note that the dimensions of ε_0 are $(TQ/M^{1/2}L^{3/2})^2 = T^2Q^2/ML^3$, so that the dimension of Q^2 in the numerator of eqn. (6.17) is canceled out to bring the right-hand side of eqn. (6.17) into compliance with the dynamical dimensions ML/T^2 of the force on the left-hand side. The point is that the coulomb of charge has been defined without regard for the dynamical relations.

In view of the fundamental role of simplicity in the construction of scientific concepts and theories, the electrodynamics and magnetohydrodynamics of the cosmos is best addressed in cgs units. The complications of the mks SI system contribute nothing to the physics, the primary effect being obfuscation of essential physical relations. We suggest as a practical matter that, if electric currents and potentials are desired in units of coulombs, amperes, and volts, one need only divide the number of cgs units of charge and current by 3×10^9 to have the desired coulombs and amperes. Thus, the charge on a proton, $e = 4.8 \times 10^{-10}$ esu, becomes 1.6×10^{-19} C. The electric potential difference in statvolts is multiplied by 3×10^2 to obtain volts. This is easy to do and avoids the conceptual damage wrought by the SI system. The reader may recall that in these conversations we have often used the volt to express potential differences and particle energies, because it connects to common laboratory experimental units.

It is particularly important in teaching the basic concepts of electricity and magnetism, and electrodynamics, to use the cgs system. Then, once the students understand the basic concepts, they can, if necessary, navigate the turbid waters of SI without serious mishap. But to start out the teaching, the concept of the electric and magnetic symmetry in cgs esu is fundamental, and the simple physics of **D** and **H** relative to **E** and **B** must not be missed.

It is amusing to note that the SI system was put forth years ago as the *practical system* of units. An excellent choice of words, for who can criticize the virtue of being "practical"? It brings to mind the classic example of Eric the Red (Eric Thorvaldson) sailing from Iceland in 982 to an ice-covered land in the western Atlantic, which he named Greenland for the fringe of meadow along the south and west sides of the ice, as would any shrewd real estate developer. Iceland, in contrast, was, and still is, a relatively verdant haven.

6.5 Systems of Units

While we are conversing on systems of units, there are several thoughts that come to mind. For instance, Newtonian gravitation is expressed in terms of the force F between two point masses m_1 and m_2 separated by a distance r, with

$$F = -G\frac{m_1 m_2}{r^2} \tag{6.21}$$

The minus sign indicates that the force is one of attraction, and G is the familiar gravitational constant, equal to 6.670×10^{-8} cm^3/g s^2. One may ask, then, why not redefine mass to be measured in units of the gravitational mass $M = G^{1/2}m$, so that

$$F = -\frac{M_1 M_2}{r^2} \tag{6.22}$$

in exact analogy to eqn. (2.1) for electric forces. Since $G^{1/2} = 2.583 \times 10^{-4}$ cgs, it follows that the unit of gravitational mass is 3.871 kg. It is our insistence on using the gram as the unit of mass that introduces the proportionality factor G into Newton's gravitational equation (2.1). We objected to the arbitrary introduction of the coulomb as the unit of electrical charge in the SI system of units, requiring, then, the otherwise unnecessary coefficient ε_0. So why not be consistent and follow the same procedure here with respect to mass so as to avoid the superfluous G? The reply is twofold. First, the principal objection to the coulomb and ε_0 was that they obscured the intrinsic symmetry of Maxwell's equations with respect to **E** and **B**. There is no such symmetry with the gravitational field, and hence no compelling reason to redefine the unit of mass. Second, the gram as the unit of mass is already thoroughly embedded in nearly every aspect of physics and daily life. The price tag on the gravitational mass is too high, considering that there is no compelling reason to adopt it.

Now, we cannot help noticing that the similar forms of eqns. (2.1) and (6.22) suggest a close analogy between electrical and gravitational fields. Indeed, there is an analogy, but with some distinct and well-known differences. In particular, there are both plus and minus electric charges, whereas matter all has the same sign for the gravitational mass, and all masses attract rather than repel each other. The Eötvös-Dicke experiment shows that anti-matter has the same gravitational properties as matter (Eötvös et al. 1922; Roll et al. 1964; Braginsky and Panov 1971).

To see how the electricity–gravitation analogy diverges, note from the comparison of eqns. (2.10) and (6.22) that the gravitational field **Z** around a point gravitational mass M is defined as

$$\mathbf{Z} = -\mathbf{e}_r \frac{M}{r^2}$$

where the minus sign is introduced so that the force on another mass M', given by $M'\mathbf{Z}$, is attractive. By analogy with the electric field, the total gravitational flux of **Z** from the mass M is obviously $4\pi M$.

Consider, then, an isolated broad uniform plane sheet of matter with surface mass density γ. In direct analogy with the electrically charged sheet in section 2.4, there is an inward directed gravitational field on either side of the sheet with intensity $2\pi\gamma$, so that a similar sheet lying parallel at a distance h is attracted with a force $2\pi\gamma^2$ per unit area. The gravitational fields of the two sheets cancel throughout the space h between them, while the fields outside have an inward directed intensity $4\pi\gamma$ from either side. This resembles the electric fields of two charged sheets with the same electrical surface charge density σ. The fields of the two charged sheets cancel between the sheets, while there is an outward electric field $E = 4\pi\sigma$ pointing away in both directions outside. The sheets repel each other with the force $2\pi\sigma^2$, and we would say that the two charged sheets are pulled apart by the net tension $E^2/4\pi - E^2/8\pi = (4\pi\sigma)^2/8\pi = 2\pi\sigma^2$ in the external field. But here the difference from the electrostatic case appears. The two sheets of matter attract rather than repel. So the external gravitational field evidently pushes the two sheets together, exerting pressure rather than tension.

Separating the attracting sheets of matter to a distance h requires the total work $W = 2\pi\gamma^2 h$, while at the same time decreasing the volume occupied by the external gravitational field $4\pi\gamma$. Doing work on the system should increase the energy of the field, but this happens at the same time that it diminishes the volume occupied by the field. The field intensity in the external volume remains fixed at $4\pi\gamma$, of course, so if the addition of energy leads to less volume of field, it follows that the energy density of the gravitational field must be negative. The work done to decrease the field volume by one unit of length is W/h, from which the energy density follows as $-Z^2/8\pi$.

When it comes to the stress in the field **Z**, it follows that there is a net pressure pushing inward on the sheets from both sides. Recalling the negative energy density, we write the stress tensor as

$$M_{ij} = +\frac{Z^2}{8\pi}\delta_{ij} - \frac{Z_i Z_j}{4\pi}$$

representing a negative isotropic pressure and a negative tension (compressive force) along the field. It is the dominance of the negative tension, i.e., pressure, in the external fields that pushes the two sheets together. The compression along the field would seem to render the field unstable to buckling, except that the isotropic negative pressure seems to stabilize it.

In summary, it is not easy to think of the gravitational field, with its negative energy, as a real physical entity in the simple direct way that we regard an electric field and a magnetic field. What we perceive as the gravitational acceleration of a free object is better thought of as a manifestation of the inward falling of space (cf. Misner et al. 1973) rather than a force field like **E** and **B**. So there are fundamental differences between the electromagnetic field and the gravitational field in spite of the identical inverse square law for both.

To get back to the discussion of systems of units, the ultimate system would be such as to reduce several fundamental constants to unity. Such a system is beloved by relativists and particle and field theorists. From the purely abstract formal point of view it is more convenient to formulate the general theory of gravitation if the gravitational constant G is set equal to one, as we have already done by introducing the gravitational mass as $M = G^{1/2}m$. To continue in this vein, the next step is to express velocity in terms of the speed of light c ($=2.9980 \times 10^{10}$ cm/s), so that the velocity v becomes $\beta = v/c$, making the speed of propagation of light equal to one. We need a third constraint to provide separate determinations of the unit of length and the unit of time, traditionally defined in terms of the size of Earth and the length of the mean solar day, respectively, and more recently in terms of the wavelength and period of particular atomic spectral lines, respectively, to provide a universal standard. For the third constraint, then, it is convenient to set Planck's constant h equal to 2π, making the more commonly employed $h/2\pi$ (h-bar) equal to one.

With the units of length, mass, and time defined in this way, let us see how the consequences show up in the classical world that we see around us. We begin by noting the dimensions of G, c, and h in terms of length l, mass m, and time t in the cgs system, with

$$[G] = \frac{l^3}{mt^2} \qquad [c] = \frac{l}{t} \qquad \left[\frac{h}{2\pi}\right] = \frac{ml^2}{t}$$

Solving these equations for l, m, and t, we have

$$t = \left(\frac{hG}{2\pi c^3}\right)^{1/2} \qquad m = \left(\frac{hc}{2\pi G}\right)^{1/2} \qquad t = \left(\frac{hG}{2\pi c^5}\right)^{1/2}$$

Given that

$$G = 6.670 \times 10^{-8}\ \mathrm{cm^3/g\ s^2}$$

$$c = 2.998 \times 10^{10}\ \mathrm{c/s}$$
$$\frac{h}{2\pi} = 1.054 \times 10^{-27}\ \mathrm{g\ cm^2/s}$$

it is readily shown that the units are

$$l = 1.615 \times 10^{-33}\ \mathrm{cm}$$
$$m = 2.177 \times 10^{-5}\ \mathrm{g}$$
$$t = 5.389 \times 10^{-44}\ \mathrm{s}$$

The unit of length is the well-known Planck length, of course, and the unit of time is just the transit time over a Planck length at the speed of light. The unit of mass is comparable to the mass of a small individual ice crystal forming a snowflake, to which we attach no fundamental significance, of course.

The unit of force, defined through Newton's law of motion as the force that increases the velocity of a unit mass by the speed of light in one unit of time, is prodigious, equal to 1.211×10^{49} dynes. The unit of density is $m/l^3 = 2\pi c^5 / Gh = 0.518 \times 10^{94}$ g/cm^3. All told, we see the advantages of the cgs system for everyday affairs, for which the cgs system was constructed in the first place.

It is not without interest to go back to inquire how it is that the Planck length, defined in the manner described above, turns out to be so small in the G, c, h system of units. Define the characteristic energy E to be mc^2, which is readily shown to be equal to $h/2\pi t$. Thus, E is recognizable as the energy quantum in terms of the characteristic frequency $\omega \equiv 1/t$. Writing the relation as $mt = h/2\pi c^2$, it is apparent that mt is small because of the smallness of $h/2\pi c^2$. Then note that $l = Gm/c^2$, representing the gravitational radius of the mass m. So with the smallness of m the gravitational radius l is doubly small because $l/m = G/c^2$ and G/c^2 is small. The extraordinary characteristic density m/l^3 is equal to c^2/Gl^2, so it is large because of both the smallness of the Planck length l and the smallness of G/c^2. That is to say, the characteristic density of a black hole increases without bound as the mass and radius of the hole decrease. In summary, then, we have the well-known fact that weak gravitation and the smallness of quanta in the macroscopic world make l so small.

Frank Wilczek (2005, 2006) provides a penetrating discussion of the choices of fundamental constants, e.g., G, c, h, or G, c, e, or h, e, m, with which one might establish convenient systems of units. He notes that ignoring the fundamental laws of physics opens the possibility for arbitrarily many different units besides length, mass, and time, with SI units as an "appalling" example, where the natural units of electric charge, defined by Coulomb's law, eqn. 2.1, are ignored and the coulomb becomes an additional unit.

6.6 Chaucer Units

When it comes to extended discussions of the relative merits of diverse systems of units, A. J. Dessler interjected the Chaucer system of units some years ago to lighten the mood. Rejecting the metric system he turned to historical precedent, with furlongs, stones, and fortnights in place of cm, g, s, or m, kg, s, respectively. For the convenience of the reader, we note here the standard conversions.

$$1 \text{ furlong} = \tfrac{1}{8} \text{ mile} = 660 \text{ feet} = 2.01 \times 10^4 \text{ cm};$$

$$1 \text{ cm} = 4.98 \times 10^{-5} \text{ furlongs}$$

$$1 \text{ stone} = 14 \text{ pound} = 6.36 \times 10^3 \text{ g}; \; 1 \text{ g} = 1.573 \times 10^{-4} \text{ stones}$$

$$1 \text{ fortnight} = 2 \text{ weeks} = 1.210 \times 10^6 \text{ s}; \; 1 \text{ s} = 0.826 \times 10^{-6} \text{ fortnights}$$

The speed of light becomes $c = 1.80 \times 10^{12}$ furlongs/fortnight. The unit of force is defined as the force that accelerates one stone at a rate of one furlong per fortnight^2, which we call the *tug*. The unit of work and energy becomes one tug furlong, and might be called the *heave*. Thus,

$$1 \text{ tug} = 0.872 \times 10^{-4} \text{ dynes}; \; 1 \text{ dyne} = 1.147 \times 10^4 \text{ tugs}$$

$$1 \text{ heave} = 1.751 \text{ ergs}; \; 1 \text{ erg} = 0.570 \text{ heaves}$$

The acceleration of gravity at the surface of Earth is, accordingly,

$$g = 980 \text{ cm/s}^2 = 7.14 \times 10^{10} \text{ furlongs/fortnight}^2$$

which might be reduced somewhat by the levity of the Chaucer system itself. One parsec is

$$1 \text{ pc} = 3.18 \times 10^{18} \text{ cm} = 1.58 \times 10^{14} \text{ furlongs;}$$
$$1 \text{ furlong} = 6.33 \times 10^{-15} \text{ pc}$$

The unit of electric charge, which might be called the *zap*, is defined in the usual way, as the charge exerting a force of one tug on an equal charge at a distance of one furlong. It turns out to be

$$1 \text{ zap} = 1.88 \times 10^{2} \text{ esu; one esu} = 5.32 \times 10^{-3} \text{ zaps}$$

The charge on an electron is then,

$$e = 4.80 \times 10^{-10} \text{ esu} = 2.56 \times 10^{-12} \text{ zaps}$$

The electric field at a distance r from a point charge q is defined as q/r^2. The magnetic field is defined by the traditional Biot relation (3.3), wherein the constant K is put equal to $1/c$. We honor the founder of the Chaucer system by naming the unit of magnetic field the dessler, just as the unit of magnetic field in the cgs system is called the gauss. With these definitions the electromagnetic equations are symmetric in **E** and **B**, and Maxwell's equations take the familiar form of eqns. (1.5) and (1.6).

Having served its purpose here, the Chaucer system merits no further comment.

7 Moving Reference Frames

7.1 Lorentz Transformations

The concept of moving reference frames has an essential role in our restless universe. The motion of the plasma throughout the cosmos involves diverse local relatively moving coordinate frames. The solar wind streams away from the Sun and rushes past the magnetospheres of Earth and the other planets. The interstellar wind blows by the heliosphere at supersonic speed, and the interstellar medium is turbulent on scales of 100 pc. Each reference frame has its own electric fields. So we have to be clear on how the electric and magnetic fields at a given point in one frame of reference are instantaneously related to those fields at the same point in other moving frames of reference.

It must also be understood that the basic laws of nature are covariant, meaning that the theoretical predictions of the details of an event are the same no matter for which moving frame of reference the calculation is carried out. So one chooses the frame of reference in which the calculation is most conveniently executed, of course.

The crucial fact is that the electric field in the local moving frame of the magnetized plasma is limited to very small values by the high electrical conductivity of the plasma, as noted in chapter 1, with the consequence that there are electric fields in other frames of reference. We will have more to say on this later, in chapter 9.

Some of the elementary concepts employed here are not fully appreciated in all fields of inquiry, so the conversation will circle around several times to provide perspective, and perhaps even to entertain, as we emphasize that the concepts should contain no unnecessary embellishments.

So the subject is the Lorentz transformation, giving the electric field $\mathbf{E}'$ and the magnetic field $\mathbf{B}'$ in the frame of reference with velocity $\mathbf{v}$ relative to another frame of reference in which the fields are $\mathbf{E}$ and $\mathbf{B}$, respectively. The essential point is that Maxwell's equations are invariant under the Lorentz transformation, fulfilling the condition that the physics of electromagnetic fields is the same in all moving reference frames. As already noted, it is sufficient for our purposes to limit the discussion to the nonrelativistic Lorentz transformations, neglecting v^2/c^2 compared to one. In that case we have

$$\mathbf{E}' = \mathbf{E} + \frac{\mathbf{v} \times \mathbf{B}}{c} \qquad \mathbf{B}' = \mathbf{B} - \frac{\mathbf{v} \times \mathbf{E}}{c} \tag{7.1}$$

again, of which we have already made considerable use in the conversation. Note that in this same nonrelativistic approximation we neglect Lorentz contraction of the coordinates (in the direction of **v**) and time dilatation. Thus, the Galilean coordinate transformations $x'_i = x_i - v_i t$, $t' = t$ apply. These transformations are presumed to be well known to the reader. Our purpose here is to examine some of their implications in the context of MHD.

7.2 Electric Fields in the Laboratory

Suppose that you are sitting in a chair in the laboratory, where $\mathbf{E} = 0$ because the laboratory is shielded from the ambient atmosphere 100 V/m, but $\mathbf{B} \neq 0$ because of the magnetic field of Earth (about 0.5 G at middle latitudes). Consider the simple question, Is there an electric field in the three-dimensional volume of the laboratory? A quick study of a free electric charge q indicates no force other than gravity exerted on the charge. So the answer is negative.

Suppose, then, that a railroad track is constructed through the laboratory and out to the horizon in both directions. A long train of cars appears in the distance, pulled along the track at a steady speed v by two 5000-kW Diesel locomotives. Each car of the train is equipped as a physics laboratory. As the train roars by you see a physicist leaning out the window of an approaching car. You shout the question, "Is there an electric field in my laboratory here?" As the car passes through the laboratory, the physicist quickly measures the force on a charge q in his railroad car laboratory and finds that the force is not zero. So he shouts, "yes," back to you at the same time that you repeat your measurement of the force on your own charge q to verify that it is still zero. It is evident from eqn. (7.1) that he measures his laboratory electric field $\mathbf{E}' = +\mathbf{v} \times \mathbf{B}/c$. To check this result you make the same inquiry of physicists in several succeeding cars. The answer is affirmative in every case.

The railroad track is then taken up and rebuilt through the laboratory in another direction. A long train of cars appears in the distance, and, as successive cars pass through the laboratory, the attending physicists measure a nonvanishing force on a charge q, shouting, "yes," to you as they depart while you check again that there is no force exerted on your own laboratory charge q.

These experimental results indicate that at any given point in space there are present simultaneously infinitely many different electric fields $\mathbf{E}' = +\mathbf{v} \times \mathbf{B}/c$, each in its own relatively moving frame of reference with velocity **v**. Each $\mathbf{E}'$ exists in its own world, i.e., reference frame, and there is no communication or interaction between different reference frames.

The detecting instrument must enter that reference frame, or moving world, to experience the corresponding electric field. There is no measurement that can be made on a charge in the laboratory frame to detect directly any of the nonvanishing **E**′ in the moving frames of reference. Equally clearly, the electric fields are there in those moving frames of reference whether we choose to look for them or not, because they are always there when we look.

Now it is sometimes stated in textbooks that the electric field experienced by an observer moving relative to a magnetic field, e.g., the magnetic field of Earth, is "induced" by the motion of the observer, implying that the electric field is caused by the presence of the moving observer and otherwise has no existence in that moving frame of reference. What a strange world is that moving reference frame! It has an electric field only if it has an observer in it. It raises the question of how the moving frame is to know whether the observer is present so that the electric field can be switched on. It is not without amusement to digress briefly to examine a parallel conundrum.

7.3 Occam's Razor and the Tree in the Forest

Consider the parallel philosophical question of whether a tree falling to the ground in the forest makes a noise in the absence of a human observer to hear the noise? If there is no observer to verify the noise, then what is the scientific basis for asserting that there is noise? One can only conjecture that there is a noise. Now the phrasing of the question clearly recognizes causality, with the novel twist that a human observer is essential to activate the causality. So with causality clearly in mind, let us play the game to see what difficulties are created by contemplating the possibility that the tree falls silently when there is no human present to record the fall. The discussion is guided by the principle of minimum embellishment and by the principle of causality.

If the falling tree is to make no sound in the absence of a human observer, then causality requires that the tree sense whether the nearest human is far enough away that the tree can safely carry out a "silent crash." This requires that the tree (living or dead) is vastly more sentient than generally supposed. So we reject the notion of a silent crash as unnecessarily complicated.

In fact, it is not possible for a tree, no matter how sentient and perceptive, to be sure of falling silently if no human is present, and with a loud crash if a human observer is available. The use of a microphone within earshot of the falling tree is sufficient to confound the notion. Suppose the microphone is connected by wire or radio to some remote observing station where the sound is recorded and then noticed sometime later by a human observer. Suppose that the microphone is switched on at intervals between periods of being turned off. So the falling tree has to be sure before emitting sound whether the microphone is on or not. Consider the dilemma, then, if the decision to switch on the microphone is made by the decay of radioactive nuclei. Perhaps one decay every couple of seconds on average. A recorded nuclear decay switches on the microphone for half a second. The microphone is 300 m from the tree, so the sound transit time from the tree to the microphone is one second. So the tree must anticipate a nuclear decay in order to decide whether to fall silently or with a crash. Once the tree hits the ground without a sound, it is too late to make a sound if the microphone proves to be on one second later. In that case a silent crash would be recorded by the human observer.

The idea that human presence is essential for the functioning of physics is absurd, of course. Present-day observations of galaxies at distances of a megaparsec or more show them to be alive and well, functioning according to the familiar laws of physics at a time when there were no humans. So we state that when there is a magnetic field but no electric field in our own laboratory frame, there is an electric field in the moving frame of reference whether there is an observer in the moving frame to sense it or not. Quite generally, if a phenomenon is detected whenever we choose to look for it, then it is always present. To assume otherwise leads to absurd situations.

7.4 Electric Field in Moving Plasma

In the context of the present concern with electric fields, we return to the point that many different $\mathbf{E}'(\mathbf{r})$ are present at any given point in the laboratory (as a consequence of $\mathbf{B}(\mathbf{r})$) in all the many different moving reference frames with nonvanishing velocity $\mathbf{v}$ relative to the laboratory, regardless of whether there is an observer in any moving frame of reference. This point of view is important when we come to the physics of MHD, wherein no single frame of reference can represent the plasma moving with nonuniform velocity $\mathbf{v}(\mathbf{r}, \mathbf{t})$. The abundance of free electrons and ions in the moving plasma erases any significant electric field in the frame of the plasma, so $\mathbf{E}'$ is set equal to zero in the local frame of reference of the swirling plasma. It follows from eqn. (7.1) that the electric field in the laboratory frame of reference is

$$\mathbf{E}(\mathbf{r}, t) = \frac{-\mathbf{v}(\mathbf{r}, t) \times \mathbf{B}(\mathbf{r}, t)}{c} \tag{7.2}$$

This electric field exists in the laboratory *because* there is no electric field in the plasma.

Now, there has been an unfortunate practice in the (**E**, **j**) paradigm literature on magnetospheric physics to assert that **E**(**r**, **t**) plays an active dynamical role in driving motions in the plasma. For instance, working in the frame of reference defined by Earth, there is an electric field $\mathbf{E}_{SW} = -\mathbf{v}_{SW} \times \mathbf{B}_{SW}/c$ in the solar wind where the plasma velocity is $\mathbf{v}_{SW}$ and the magnetic field is $\mathbf{B}_{SW}$. It is sometimes asserted that $\mathbf{E}_{SW}$ actively penetrates from the wind into the magnetosphere, thereby setting the plasma in the geotail into motion approaching the electric drift velocity

$$\mathbf{v}_D = c\frac{\mathbf{B} \times (\mathbf{v}_{SW} \times \mathbf{B}_{SW})}{B^2}$$

where now **B** is the geomagnetic field. Note that this motion can be in most any direction, depending on the relative orientation of **B** and $\mathbf{B}_{SW}$. The antisolar streaming wind could just as well drive a solar-directed convection, contrary to the observed antisolar motion of the outer layers of the geomagnetic field. However, it is alleged that this is the basic cause of magnetospheric convection. A cross section of the geotail is sometimes drawn with the magnetotail between two wide parallel condenser plates, between which there is an applied potential difference in excess of 10^4 V in the direction to give the actual convection. But if the electric field were a driving force, then in which frame of reference are we to use the electric field for computing the driving of the plasma? All those electric fields in the many different moving frames of reference are there, eagerly waiting to be exploited. There is, of course, no comparable electric field penetrating from the magnetosphere outward to affect the wind, so it appears that the wind suffers no gain or loss of momentum. On the other hand, if the calculation is carried out in the frame of the wind, then there is no electric field in the wind to drive magnetospheric convection, but rather an electric field $+\mathbf{v}_{SW} \times \mathbf{B}/c$ in the magnetosphere that, by the same principle, actively penetrates into the solar wind causing the electric drift velocity $(\mathbf{v}_{SW} \times \mathbf{B}) \times \mathbf{B}_{SW}/B^2$ of the solar wind plasma relative to the frame of reference of the solar wind. The principle of covariance is violated, and whatever happened to conservation of momentum?

This author suggests that the electric field **E**′ in the frame of the plasma is negligible, based on the considerations leading to eqn. (1.8). Another prejudice of the author is that Newton's equation of motion is the proper venue for the discussion of driven motions. Yet, when we come to Newton's momentum equation in chapter 8, we find no such driving effect introduced by electric fields. Indeed, it is obvious from eqn. (7.2) that the stresses, $E^2/8\pi$, in the electric field are small $O(v^2/c^2)$ compared to the stresses, $B^2/8\pi$, in the magnetic field. That is to say, the stresses in the electric field are small to the same order as time dilatation and Lorentz contraction, both of which we neglect to very good approximation. So the electric stresses are to be neglected. Vasyliunas (2001, 2005a, b) has investigated in detail the dynamical role of **E**, showing how **E** is shaped by the plasmas motion, rather than vice versa.

In contrast, in the **E**, **j** paradigm the force $\pm e\mathbf{E}$ on an individual ion or electron looms large in the generalized Ohm's law and is considered to be an important dynamical effect. The net force on the plasma is very small, of course, because of the electrical neutrality of the plasma, so it is evident that $\pm e\mathbf{E}$ has no role in the large-scale dynamics in spite of its prominence in the **E**, **j** paradigm. This is just one more illustration of how the **E**, **j** paradigm has such difficulty in coming to grips with the time-dependent dynamics of the plasma.

But what, then, of the very small $\mathbf{E}'$ in the frame of the plasma itself, necessary to maintain the electric current required by Ampere's law? What electric drift velocity might it produce? Alfvèn and Fälthammer (1963) suggested that the electric drift velocity driven by $\mathbf{E}'$ causes the bright coronal filaments to shrink, thereby explaining their remarkable threadlike appearance. The idea is easily illustrated by considering a twisted flux tube extending along the z axis, with field components $B_\varphi(\varpi)$, $B_z(\varpi)$, where ϖ represents radial distance from the z axis, and φ is the azimuthal angle measured around the z axis. The current along the twisted flux bundle is given by

$$j_z = \frac{c}{4\pi\varpi}\frac{d\varpi B_\varphi}{d\varpi}$$

The plasma has some slight resistivity so there must be the longitudinal electric field

$$E_z' = \frac{j_z}{\sigma}$$

in the frame of the plasma, where σ is the electrical conductivity. The electric drift velocity in the frame of the plasma $c\mathbf{E}' \times \mathbf{B}/B^2$ is, then, cE_zB_φ/B^2 directed radially inward toward the axis of the twisted bundle. So the radius of the current-carrying bundle should shrink slowly with the passage of time.

Unfortunately, there is no such effect. The electric force $\pm e\mathbf{E}'$ exerted on each ion and electron is exactly canceled in the mean by the frictional drag (resistivity). The electric field $\mathbf{E}'$ is nonvanishing precisely because there is frictional opposition to the steady electron conduction velocity.

The charged particles respond to the total force **F** exerted on them and accelerate accordingly. In a steady flow of current there is no total force so there is no radial contraction of the twisted flux bundle. In this connection note that an electric charge q experiencing a force **F** perpendicular to a magnetic field **B** drifts with the velocity

$$\mathbf{u}_D = c\frac{\mathbf{F} \times \mathbf{B}}{qB^2}$$

so that the Lorentz force $q\mathbf{u}_D \times \mathbf{B}/c$ exactly cancels the force **F**. There is no inward drift of the ions and electrons, and the remarkable fineness of the coronal filaments has some other origin.

7.5 Net Charge in a Swirling Plasma

Having emphasized that there are different electric fields in each of the infinitely many relatively moving frames of reference, the obvious question is what are we to think of the electric charge density δ in eqn. (1.6), which is different in each different moving frame of reference. With

$$\nabla \cdot \mathbf{E} = 4\pi\delta$$

it follows with the aid of eqn. (7.2) that

$$\delta = \frac{1}{4\pi}\left(\frac{\mathbf{v}}{c}\cdot\nabla \times \mathbf{B} - \mathbf{B}\cdot\nabla \times \frac{\mathbf{v}}{c}\right) \qquad (7.3)$$

Now, $\nabla \times \mathbf{v}$ is invariant from one reference frame to the next, so there is no particular puzzle concerning that term. The interesting effect is the first term on the right-hand side, which is different in each moving frame. So suppose that **v** is uniform, so that

$$\delta = \frac{1}{4\pi}\frac{\mathbf{v}}{c}\cdot\nabla \times \mathbf{B}$$

$$= \frac{\mathbf{v}\cdot\mathbf{j}}{c^2} \qquad (7.4)$$

upon using Ampere's law, eqn. (3.11). Then, if **u** represents the mean conduction velocity of the electrons relative to the ions, with N electrons/cm^3, it follows that $\mathbf{j} = -Ne\mathbf{u}$, and

$$\delta = -\frac{Ne}{c^2}\mathbf{v}\cdot\mathbf{u} \qquad (7.5)$$

It is this small charge density that is associated with the radial electric field that produces the Poynting vector, representing the energy flow associated with the electric current.

The electrostatic force per unit volume is

$$\begin{aligned}\mathbf{F} &= \delta\mathbf{E} \\ &= \frac{e(\mathbf{v}\cdot\mathbf{j})(\mathbf{v}\times\mathbf{B})}{16\pi^2c^3}\end{aligned} \tag{7.6}$$

This is clearly small $O(v^2/c^2)$ and quite negligible, compared to the Lorentz force $\mathbf{j} \times \mathbf{B}/c$. The question lies with the charge density δ, which, in principle, is a measurable physical quantity no matter how small. Yet it is different in each differently moving frame of reference. Certainly the total number of stable particles is the same in all reference frames. So how can the difference in the number density of the electrons and ions vary from one reference frame to another? The well-known answer is simply that the particle density involves the scale of the volume in each different reference frame, so the number densities can be compared only if there exists a reference frame in which the electrons and ions are both at rest. However, in the presence of a nonvanishing current $\mathbf{j}$ the motions of the electrons and ions differ by the conduction velocity $\mathbf{u}$, so that the densities cannot be compared directly.

For simplicity suppose that $\mathbf{v}$ is parallel to $\mathbf{j}$ and suppose that the ions are singly charged and have the same mass as the electrons. Then, if $\mathbf{v}$ represents the mean bulk motion of the plasma, giving equal weight to ions and electrons, it follows that the mean bulk motion of the electrons is $\mathbf{v} + \mathbf{u}/2$ and the mean bulk velocity of the ions is $\mathbf{v}-\mathbf{u}/2$. With the number density N of the electrons and ions each in its own moving frame, the Lorentz contraction provides the electron density

$$\begin{aligned}N_e &= \frac{N}{\left[1 - \frac{(v + u/2)^2}{c^2}\right]^{1/2}} \\ &\cong N\left[1 + \frac{(v + u/2)^2}{2c^2} + \cdots\right]\end{aligned}$$

and the ion number density is

$$\begin{aligned}N_i &= \frac{N}{\left[1 - \frac{(v - u/2)^2}{c^2}\right]^{1/2}} \\ &\cong N\left[1 + \frac{(v + u/2)^2}{2c^2} + \cdots\right]\end{aligned}$$

The charge density in the laboratory is, then,

$$\begin{aligned}\delta &= e(N_i - N_e)\\ &= -\frac{eN\mathbf{v}\cdot\mathbf{u}}{c^2}\\ &= \frac{\mathbf{v}\cdot\mathbf{j}}{c^2}\end{aligned}$$

to lowest order. This is, of course, precisely the net charge density given by eqn. (7.5). So total charge is invariant under Lorentz transformation, but the charge density is not, as a consequence of Lorentz contraction. The charge density $\delta = \nabla \cdot \mathbf{E}/4\pi$ is automatically supplied by Lorentz contraction. The actual case, in which the ion mass M is large compared to the electron mass m, is worked out in Appendix D.

The essential point, then, is that, neglecting resistivity, the electric field $\mathbf{E}'$ vanishes in the local Lagrangian element of volume of plasma, moving with the swirling velocity $\mathbf{v}(\mathbf{r}, \mathbf{t})$ of the plasma. The electric field in the laboratory is, therefore, $\mathbf{E}(\mathbf{r}, t) = -\mathbf{v}(\mathbf{r}, t) \times \mathbf{B}(\mathbf{r}, t)/c$, and is responsible for no dynamical effects. In some other frame, with velocity $\mathbf{V}$ relative to the laboratory, there is, of course, the electric field $\mathbf{E}_V$, given by

$$\mathbf{E}_V(\mathbf{r}, t) = \frac{-[\mathbf{v}(\mathbf{r}, t) - \mathbf{V}] \times \mathbf{B}(\mathbf{r}, t)}{c}$$

The essential point is that the dynamics of the bulk motion of the plasma can be carried out in any desired frame of reference, with arbitrary velocity $\mathbf{V}$ relative to the laboratory, and there is a different electric field with each choice of $\mathbf{V}$. The difference of $\mathbf{E}_V(\mathbf{r}, t)$ from the electric field in the laboratory is $+\mathbf{V} \times \mathbf{B}(\mathbf{r}, t)/c$, the form of which can vary qualitatively from one frame to the next, particularly if $\mathbf{B}(\mathbf{r}, t)$ is not simple in form. So the electric field cannot be incorporated into the dynamics of the plasma motion $\mathbf{v}$ without creating contradiction. This nonparticipation requirement is satisfied by the fact that the electric field stress and energy are small to second order in v/c compared to the magnetic stress and energy, and hence quite negligible.

8 Hydrodynamics

8.1 Basic Considerations

The magnetic fields throughout the cosmos are embedded in plasma so that the dynamics of the magnetic fields involves the mechanics of the plasma motion influenced by the stresses in the magnetic field. So if we are to understand the dynamics of the magnetic field, we must understand the dynamical properties of the large-scale bulk motion of the plasma. The large-scale dynamics of a plasma is described by the equations of hydrodynamics, with conservation of the plasma number density N described by

$$\frac{\partial N}{\partial t} + \nabla \cdot (N\mathbf{v}) = 0 \tag{8.1}$$

and conservation of momentum described by

$$NM\left[\frac{\partial \mathbf{v}}{\partial t} + (\mathbf{v} \cdot \nabla)\mathbf{v}\right] = -\nabla p + \mathbf{F} \tag{8.2}$$

where M is the atomic mass, p the plasma pressure, $\mathbf{F}$ the applied force per unit volume, and $\mathbf{v}$ the bulk velocity.

Now, there is a persistent folklore to the effect that hydrodynamics is not applicable in the absence of interparticle collisions. It is stated that pressure cannot be defined, the thermal velocity distribution may not be Maxwellian, etc. What is more, it is asserted that MHD does not apply because there is no applicable Ohm's law, etc. The standard textbooks on hydrodynamics and MHD generally do not address these questions directly, focusing attention on the formal mathematical derivation of the equations from the Boltzmann equation, etc. The approach that we take here is to do no unnecessary mathematics, but instead to be guided by the elementary physical principles. Too often the formal mathematical approach makes restrictive assumptions essential for pursuing the mathematics, but not required for the physics. Our purpose is to show that the familiar hydrodynamic equations are the unavoidable consequence of the simplest physical principles, applying with or without interparticle collisions or thermal isotropy. The conservation of particles and the conservation of momentum are unaffected by collisions because collisions also conserve particles and momentum. Thus, eqns. (8.1) and (8.2) are inescapable regardless of the presence or absence of collisions. The rate of inteparticle collisions appears only when we come to compute the pressure, which becomes a tensor p_{ij} when collisions do not dominate. In the presence of a magnetic field, the pressures $p_\perp$ and $p_\parallel$ perpendicular and parallel to **B** may differ. The cyclotron motion of the electrons and ions around the magnetic field guarantees isotropy of the large-scale pressure $P_\perp$ in the two dimensions perpendicular to the field.

For the extreme case of a collisionless plasma it is customary to deduce eqns. (8.1) and (8.2) by computing the velocity moments of the Boltzmann equation for the velocity distribution function $f(\mathbf{r}, \mathbf{u}, t)$ of the individual particle velocities **u**. The velocity distribution function is integrated out of the system, obtaining eqns. (8.1) and (8.2) for any velocity distribution, Maxwellian or otherwise. However, eqns. (8.1) and (8.2) follow on a much simpler basis, which makes clear their universal applicability. The essential point is simply that the time rate of change of the density of a conserved quantity is equal to the negative divergence of the flux density of that quantity. Both eqns. (8.1) and (8.2) are constructed from nothing more than that simple fact, with an applied external force F_i subsequently added to the right-hand side of the momentum equation.

Consider, then, what conditions must be met if the dynamics of a collisionless gas is to be described by equations of the form of (8.1) and (8.2). Obviously, there must be enough particles (atoms, ions, electrons) to give a sufficiently precise statistical determination of the local number density N. We are concerned with the dynamics of the large-scale bulk flow (characteristic scale Λ) described by the partial differential equations of hydrodynamics. Look upon these partial differential equations as difference equations with a small grid spacing $\lambda (\lambda << \Lambda)$. Then consider the expected number of particles $N\lambda^3$ in each basic cell. If the difference equations are to produce a useful mathematical approximation, it is clear that λ must be sufficiently small compared to Λ. It is also clear that there must be enough particles $N\lambda^3$ in each basic cell so as to provide a statistically well-defined density N, because the difference equations work with the mean density, ignoring the statistical fluctuations of the order of $(N\lambda^3)^{1/2}$, which cause the actual physical system to drift away from the exact mathematical solution of the difference, or differential, equations. The need for statistical precision of the mean density must be balanced against the required smallness of λ compared to Λ. The relative magnitude of these two requirements depends on the precision of the initial conditions and how far the dynamics is to be carried forward in time.

In most astronomical settings the statistical requirement on $N\lambda^3$ is readily satisfied. For instance, it seems adequate for most purposes to have $\lambda = 10^{-3}\ \Lambda$. Consider the large-scale interaction of the solar wind ($N \approx$ 5 ions/cm^3) with the magnetosphere of Earth ($\Lambda \approx 10^9$cm). It follows that

$\lambda = 10^6$ cm and $N\lambda^3 = 5 \times 10^{18}$ ions. The statistical uncertainty in N is less than one part in 10^9, and is far more precise than necessary. The scale λ could be much smaller. However, it must be appreciated that we do not intend to take on the complications of the detailed cyclotron motion of the ions as they impact the geomagnetic field (cf. Parker 1967). So for purely hydrodynamic considerations we consider the bulk motion only over scales that are large compared to the ion cyclotron radius, etc.

Even the tenuous relativistic cosmic-ray gas ($N \approx 10^{-10}/\text{cm}^3$) has enough particles to provide a statistically well-defined number density on a scale of 10^7 cm, with $N\lambda = 10^{11}$ particles. In fact, the cyclotron radii of the cosmic ray particles are typically 10^{11}cm or more, so hydrodynamics of the cosmic-ray gas is inapplicable on scales of 10^{11}cm or less, or about twice the Earth–Moon distance. On scales of 10^{12} cm (0.07 AU) and larger the cosmic ray gas behaves as a fluid. The point is simply that for the *large–scale* bulk motions of plasmas there is generally no problem with having enough particles to provide a well-defined N.

8.2 Derivation of the HD Equations

Consider how the two-scale concept, with $\lambda << \Lambda$, works out formally. If the number density N is well defined statistically, then so are the mean bulk velocity, the effective kinetic temperature of the ions and the electrons, and the pressure. To proceed, denote the velocity of an individual particle by u_i or $\mathbf{u}$, the mean bulk velocity by v_i or $\mathbf{v}$, and thermal velocity, defined as the velocity of the particle relative to the mean, by w_i or $\mathbf{w}$. Then

$$\mathbf{u} = \mathbf{v} + \mathbf{w} \tag{8.3}$$

The mean bulk velocity is obtained by summing over all particles in the small volume $V = \lambda^3$. It is sufficient for present purposes of establishing HD to consider only a single type of particle—a one-fluid theory—with generic mass M, presumably the ions. We are ignoring the fact that the mean velocity of the electrons differs slightly from the mean velocity of the ions in the presence of an electric current. The current is determined through Ampere's law, eqn. (3.1), by the deformation of the magnetic field in the dynamical interaction between the plasma and field. For the record, one finds electron conduction velocities of meters/sec in the low electron densities in the solar photosphere and in the solar corona, less in the solar wind, and microscopic (10^{-6}–10^{-7} cm/s) on galactic scales. This question comes up again in chapter 10, where it is dealt with in more detail in a partially ionized gas (see also Vasyliunas and Song 2005).

The mean velocity v_i in the volume V is given by the sum Σ over all particles in V,

$$\begin{aligned} v_i &= \frac{1}{NV}\sum u_i \\ &= \frac{1}{NV}\sum (v_i + w_i) \\ &= v_i + \frac{1}{NV}\sum w_i \end{aligned}$$

from which it follows that

$$\sum w_i = 0 \tag{8.4}$$

because the thermal motions are defined to have no mean.

The mean particle flux is Nv_i, so conservation of particles requires that the time rate of change of the particle density N is given by the negative divergence of the particle flux,

$$\frac{\partial N}{\partial t} = -\frac{\partial}{\partial x_j} N v_j \tag{8.5}$$

which is just eqn. (8.1) again.

Similarly, the momentum density NMv_i follows from

$$\frac{1}{V}\sum M u_i = NMv_i$$

The j component of the flux of momentum density in the i direction is

$$\begin{aligned} \frac{1}{V}\sum M u_i u_j &= \frac{1}{V}\sum M(v_i + w_i)(v_j + w_j) \\ &= \frac{1}{V}\sum M(v_i v_j + w_i w_j) \\ &= NMv_i v_j + \frac{1}{V}\sum M w_i w_j \end{aligned}$$

upon using eqn. (8.4). The second term on the right-hand side represents the flux density of momentum transported by the thermal motions alone.

That is what is meant by pressure. The thermal motions may not be statistically isotropic, so the pressure is a tensor, defined as

$$p_{ij} = \frac{1}{V}\sum M w_i w_j \tag{8.7}$$

and the flux of momentum density becomes

$$\frac{1}{V}\sum M u_i u_j = NM v_i v_j + p_{ij} \tag{8.8}$$

Pressure has nothing to do with collisions between particles. It is simply the momentum flux density transported by the thermal motions, just as the Reynolds stress,

$$R_{ij} = NM v_i v_j \tag{8.9}$$

represents the momentum flux density transported by the mean bulk velocity. Note, then, that $p_{ij} = p_{ji}$, just as $R_{ij} = R_{ji}$.

The time rate of change of the momentum density is equal to the negative divergence of the flux of momentum density, yielding

$$\frac{\partial}{\partial t} NM v_i = -\frac{\partial}{\partial x_j}\left(NM v_i v_j + p_{ij}\right) \tag{8.10}$$

This is the statement of momentum conservation in the statistically well-defined fluid.

It is obvious that a bulk force F_i per unit volume, e.g., gravity or Maxwell stress, applied to the fluid has no direct effect on the thermal motions and is merely added to the right-hand side of eqn. (8.10), giving

$$\frac{\partial}{\partial t} NM v_i = -\frac{\partial}{\partial x_j}\left(NM v_i v_j + p_{ij}\right) + F_i \tag{8.11}$$

Now multiply eqn. (8.5) by v_i and subtract from eqn. (8.11), providing the momentum equation in the familiar form

$$NM\left(\frac{\partial v_i}{\partial t} + v_j \frac{\partial v_i}{\partial x_j}\right) = -\frac{\partial p_{ij}}{\partial x_j} + F_i \tag{8.12}$$

It should be remarked that this simple development of the HD equations was carried out with the collisionless plasma in mind. It is evident, however, that the same equations could be constructed for a plasma with interparticle collisions, because collisions conserve both particles and the total momentum of colliding particles. In fact, about the only assumption that might break down in the dense gas is the implicit idea that the diameters of the particles are small compared to the interparticle distances. Van der Waal's equation of state recognizes the effect of particle diameters and long-range interparticle forces. However, this seems not to be significant in the plasmas in the cosmos. Even in the dense atmosphere of Earth, the air at 300 K and a pressure of 10^6 dynes/cm^2 has a number density N of about 10^{19}/cm^3. The distance between molecules is of the order of 4×10^{-7}cm, or about 10 times the molecular dimension. Thus, air has a density of about 10^{-3} of the density of the close packed molecules of liquid nitrogen or oxygen or water.

8.3 The Pressure Tensor

Consider the calculation of the pressure tensor p_{ij}. Collisions enter the picture at this point, pushing the thermal motions toward statistical isotropy. So again we use the collisionless plasma to establish the basic relations and principles, introducing the scattering by collisions once the basic dynamics is in hand. Note, then, that in the absence of collisions and in the absence of external forces, the individual particles move freely in straight lines, conserving the momentum and energy of each velocity component, u_i. The question is how an inhomogeneous distribution of these freely moving particles evolves, providing the time variation of p_{ij}. To begin, then, note that the diagonal terms of p_{ij}, viz. NMw_x^2, NMw_y^2, NMw_z^2 represent (twice) the kinetic energy of the thermal motions in each direction. This suggests that we should begin with the kinetic tensor $V^{-1}\Sigma Mu_iu_j$, representing the momentum flux density. The tensor $V^{-1}\Sigma Mu_iu_j$ changes with the passage of time because particles individually pass in and out of the volume V over which the particles are summed. Hence, the time rate of change of the tensor $V^{-1}\Sigma Mu_iu_j$ is equal to the negative divergence of the flux density $V^{-1}\Sigma M(u_iu_j)u_k$ of the tensor $V^{-1}\Sigma Mu_iu_j$,

$$\frac{\partial}{\partial t}\frac{1}{V}\sum Mu_iu_j = -\frac{\partial}{\partial x_k}\frac{1}{V}\sum Mu_iu_ju_k \tag{8.13}$$

This becomes

$$\frac{\partial}{\partial t}NMv_iv_j + \frac{\partial}{\partial t}\frac{1}{V}\sum NMw_iw_j$$

$$= -\frac{\partial}{\partial x_k}NMv_iv_jv_k - \frac{\partial}{\partial x_k}\left(v_k\frac{1}{V}\sum Mw_iw_j + v_j\frac{1}{V}\sum Mw_iw_k\right.$$

$$\left. + v_i\frac{1}{V}\sum Mw_jw_k + \frac{1}{V}\sum Mw_iw_jw_k\right)$$

At this point we introduce the heat flow tensor T_{ijk}, defined a

$$T_{ijk} = \frac{1}{V}\sum Mw_iw_jw_k \tag{8.14}$$

and representing the transport of kinetic energy by the thermal motions. The terms first order in w_i have dropped out because they sum to zero, as indicated by eqn. (8.4). The result is

$$\frac{\partial}{\partial t}NMv_iv_j + \frac{\partial}{\partial t}\frac{1}{V}\sum Mw_iw_j = -\frac{\partial}{\partial x_k}\left(NMv_iv_jv_k + v_kp_{ij}\right.$$

$$\left. + v_jp_{ik} + v_ip_{jk} + T_{ijk}\right)$$

The next step is to eliminate the time derivative of NMv_iv_j from the left-hand side, writing the time derivative as $NMv_i\partial v_j/\partial t + v_j\partial NMv_i/\partial t$ and using eqns. (8.10) and (8.12). The final result is the expression

$$\frac{\partial p_{ij}}{\partial t} + v_k\frac{\partial p_{ij}}{\partial x_k} = -p_{ij}\frac{\partial v_k}{\partial x_k} - p_{ik}\frac{\partial v_j}{\partial x_k} - p_{kj}\frac{\partial v_i}{\partial x_k} - \frac{\partial T_{ijk}}{\partial x_k}$$

for the Lagrangian derivative of the pressure tensor p_{ij}.

This equation includes thermal conduction in the term $\partial T_{ijk}/\partial x_k$, and, of course, there may be other sources of heat, such as the dissipation of waves, or a myriad of nanoflares, as in the solar wind and in the active solar corona, respectively, which lie outside the local pressure equation. So we introduce a generic source term S_{ij} on the right-hand side, in the same ad hoc way that we included the externally applied force F_i on the right-hand side of eqns. (8.11) and (8.12). Thus,

$$\frac{\partial p_{ij}}{\partial t} + v_k\frac{\partial p_{ij}}{\partial x_k} = -p_{ij}\frac{\partial v_k}{\partial x_k} - p_{ik}\frac{\partial v_j}{\partial x_k} - p_{kj}\frac{\partial v_i}{\partial x_k} - \frac{\partial T_{ijk}}{\partial x_k} + S_{ij} \tag{8.15}$$

The next formal mathematical step would be to work out the equation for the heat flow tensor T_{ijk}. However, it is well known that the equation for $\partial T_{ijk}/\partial t$ involves the fourth-order tensor W_{ijkl}, given by $(1/V)\Sigma M w_i w_j w_k w_l$. Then the equation for W_{ijkl} involves the fifth-order tensor X_{ijklm}, given by $(1/V)\Sigma M w_i w_j w_k w_l w_m$, ad infinitum. Over the years various investigators have developed a variety of schemes for terminating this infinite system of equations, each scheme speaking to its own physical circumstances. However, in astrophysical settings we usually do not know either the initial conditions or the boundary conditions on W_{ijkl}, X_{ijklm}, etc., and we do not have an observational grasp on the diverse components of T_{ijk}. The complex mathematical procedures provide little or no enlightenment on these questions. The procedures serve more to overshadow and obscure the limited state of knowledge of the actual heat flow. So we suggest that the heat conduction may as well be treated by simpler means in most cases. Heat flow can usually be estimated from simple physical considerations on an effective, and perhaps saturated, thermal conductivity. It must also be borne in mind that the heat source S_{ij} in and around the atmosphere of a star and its stellar wind is at least as important as heat conduction, and the precise nature of S_{ij} is usually not known at all. As an illustrative example, note that the solar wind crossing the orbit of Earth is expanding principally in the two transverse directions, with the radial velocity of the wind essentially uniform. Therefore, we would expect the transverse temperature to be substantially less than the radial temperature. However, in situ observations in the wind show that there are times when the transverse ion temperature equals or exceeds the radial temperature. This indicates a vigorous heat input to the transverse thermal motions of the ions, presumably the result of transverse plasma waves near the ion cyclotron frequency somewhere inside the orbit of Earth. In contrast, the electron temperatures at the orbit of Earth seem to be controlled largely by heat conduction from the hot corona back at the Sun, so their temperature is usually below the ion temperature and controlled by heat conduction alone. The overall dynamical picture is then limited by the empirical facts. There should be an approximate "impedance match" between the detailed knowledge of the physical situation and the complexity of the mathematics employed in modeling that situation. This modus operandi is in conformity with the advice of H. P. Robertson some 60 years ago: "When the mathematics gets too difficult, it is time to stop and think about the physics."

So, to review the basic physics of the thermal variations in elementary flows, note first that an incompressible flow of a uniform fluid (for which $p_{ij} = p\delta_{ij}$ and the divergence condition $\nabla \cdot \mathbf{v} = 0$ replaces eqn. (8.15) for the time variation of the pressure p) encompasses hydrodynamic turbulence. That subject alone has engaged the minds of some the world's best physicists and mathematicians since the days of Reynolds (cf. Reynolds 1883, 1895; Lamb 1932; Goldstein 1938; Kolmogoroff 1941a,b, 1962; Landau and Lifschitz 1959; Batchelor 1967, 1971; Kraichnan 1974; Kraichnan and Montgomery 1980; McComb 1990; Saffman 1992 and references therein) and has advanced far beyond the scope of these simple conversations. In accordance with Robertson's dictum, it is customary to handle turbulence in the cosmos with the physical concepts of eddies and the associated mixing length, usually ignoring such refinements as compressibility, statistical anisotropy, and intermittency, but nonetheless providing useful insights into the gross dynamical features of a turbulent fluid.

Now, these conversations are directed to the HD of a plasma, admitting of compressibility and thermal anisotropy in addition to classical incompressible HD turbulence. This adds considerably to the theoretical complexity, even without introducing the magnetic field to give MHD. So we first have a look at the dynamical ramifications of eqn. (8.15) to understand how the pressure tensor p_{ij} is manipulated by a nonuniform bulk flow **v**. We examine the effects of a uniform but anisotropic dilatation and of a uniform shear that we may appreciate how these effects proceed with generally nonuniform velocity fields. This elementary exploration of the physics is carried through into the next chapter, culminating in the Chew-Goldberger-Low double-adiabatic (invariant) treatment of the pressure anisotropy in the presence of a nonuniform magnetic field. This physical approach does not have the elegance of formal mathematical development from the theoretical foundations, of course, but it provides sufficient guidance to investigate the dynamical problems thrust upon us by the observations and to point the way for others to follow bearing greater loads of mathematical machinery. The formal mathematical approach is available in various monographs (cf. Wu 1966) for comparison with the results of these conversations.

8.4 Pressure Variations in Uniform Dilatations

Consider the uniform one-dimensional expansion in the $i = 1$ direction, with $\partial v_1/\partial x_1 = 1/\tau_1$ and all other components of $\partial v_i/\partial x_j$ put equal to zero. In the absence of large-scale shear, we presume that the off-diagonal terms of p_{ij} are negligible. It follows from eqn. (8.15) that the diagonal terms are given by

$$\frac{dp_{11}}{dt} = -3\frac{p_{11}}{\tau_1} \qquad \frac{dp_{22}}{dt} = -\frac{p_{22}}{\tau_1} \qquad \frac{dp_{33}}{dt} = -\frac{p_{33}}{\tau_1} \tag{8.16}$$

With

$$\frac{dN}{dt} = -\frac{N}{\tau_1} \tag{8.17}$$

from eqn. (8.5), it follows that

$$p_{11}(N) = p_{11}(N_0)\left(\frac{N}{N_0}\right)^3 \qquad p_{22}(N) = p_{22}(N_0)\left(\frac{N}{N_0}\right)$$

$$p_{33}(N) = p_{33}(N_0)\left(\frac{N}{N_0}\right) \tag{8.18}$$

within any moving element of plasma. This is the familiar one-dimensional adiabatic expansion.

For expansion in three dimensions let

$$\frac{\partial v_1}{\partial x_1} = \frac{1}{\tau_1} \qquad \frac{\partial v_2}{\partial x_2} = \frac{1}{\tau_2} \qquad \frac{\partial v_3}{\partial x_3} = \frac{1}{\tau_3}$$

Then eqn. (8.15) yields

$$\frac{dp_{11}}{dt} = -p_{11}\left(\frac{3}{\tau_1} + \frac{1}{\tau_2} + \frac{1}{\tau_3}\right) \qquad \frac{dp_{22}}{dt} = -p_{22}\left(\frac{1}{\tau_1} + \frac{3}{\tau_2} + \frac{1}{\tau_3}\right)$$

$$\frac{dp_{33}}{dt} = -p_{33}\left(\frac{1}{\tau_1} + \frac{1}{\tau_2} + \frac{3}{\tau_3}\right) \tag{8.20}$$

with

$$\frac{dN}{dt} = -N\left(\frac{1}{\tau_1} + \frac{1}{\tau_2} + \frac{1}{\tau_3}\right) \tag{8.21}$$

Thus, the pressure in a moving element of plasma varies as

$$p_{11}(N) = p_{11}(N_0)\left(\frac{N}{N_0}\right)^{1+2\alpha(1)_1} \qquad p_{22}(N) = p_{22}(N_0)\left(\frac{N}{N_0}\right)^{1+2\alpha(2)}$$

$$p_{33}(N) = p_{33}(N_0)\left(\frac{N}{N_0}\right)^{1+2\alpha(3)} \tag{8.22}$$

with the density N, where

$$\alpha(n) = \frac{1/\tau_n}{1/\tau_1 + 1/\tau_2 + 1/\tau_3} \tag{8.23}$$

This is the general form of adiabatic cooling in a collisionless plasma. An isotropic expansion ($\tau_1 = \tau_2 = \tau_3$) causes the pressure to vary in proportion to $N^{5/3}$. A two-dimensional expansion ($\tau_1 = \tau_2 \neq 0$, $\tau_3 = \infty$) causes p_{11} and p_{22} to vary in proportion to N^2, while p_{33} varies only as N.

It is easy to understand these results in terms of the free motion of the individual particles. It is sufficient to treat the one-dimensional uniform expansion, with $v_1 = x_1/\tau_1$, i.e., $\partial v_1/\partial x_1 = 1/\tau_1 > 0$. Consider a particle moving in the x_1 directionn with constant velocity $u_1 = v_1 + w_1$. Then, since $du_1/dt = 0$, it follows that

$$\frac{dw_1}{dt} = -\frac{dv_1}{dt} \tag{8.24}$$

$$= -\frac{dv_1}{dx_1}\frac{dx_1}{dt} \tag{8.25}$$

$$= -\frac{1}{\tau_1}w_1 \tag{8.26}$$

as the particle crosses $x_1 = 0$, where $v_1 = 0$. Thus, the thermal velocity declines with the passage of time. The density is also decreasing, in the manner described by eqn. (8.17). When eqns. (8.17) and (8.24) are combined, the result is

$$\frac{dw_1}{dN} = \frac{w_1}{N} \tag{8.27}$$

indicating that w_1 varies in direct proportion to N. So the individual particle does not lose energy in the dilatation $\partial v_1/\partial x_1$, but the division of u_1 between the bulk velocity v_1 and the thermal velocity w_1 changes so that w_1 diminishes. Since $p_{11}$is proportional to $N{w_1}^2$, it follows that p_{11} is proportional to N^3 . In the same situation, p_{22} and p_{33} decline only in proportion to N because w_{22} and w_{33} are unaffected.

A more conventional approach to adiabatic cooling is to introduce plane reflecting walls at $x_1 = \pm L(t)$ moving with the fluid, so that $dL/dt = L/\tau_1$. Working in the limit that $L/\tau_1 << w_1$, we consider a particle bouncing back and forth between the receding walls. The particle collides with a wall at time intervals

$$\Delta t = \frac{2L}{w_1} \tag{8.28}$$

losing velocity

$$\Delta w_1 = -\frac{2dL}{dt} = -\frac{2L}{\tau_1} \tag{8.29}$$

with each wall collision. During the time interval Δt the length L increases by

$$\Delta L = \frac{L\Delta t}{\tau_1} \tag{8.30}$$

It follows that

$$\frac{\Delta w_1}{w_1} = -\frac{\Delta L}{L}$$

Integration provides the well-known longitudinal invariant,

$$w_1(t)L(t) = w_1(0)L(0) \tag{8.31}$$

Then, since

$$N(t)L(t) = N(0)L(0) \tag{8.32}$$

it follows that $w_1(t)$ is proportional to $N(t)$, as concluded from eqn. (8.27).

8.5 Shear Flow

With these remarks on the diagonal terms of p_{ij}, consider the behavior of the off-diagonal terms in the presence of a simple laminar shear flow. Suppose that $\partial v_1/\partial x_2 = 1/\tau_{12}$ with all the other components of $\partial v_i/\partial x_j$ equal to zero. It follows from eqn. (8.15) that

$$\frac{dp_{11}}{dt} = -\frac{2p_{12}}{\tau_{12}}, \quad \frac{dp_{22}}{dt} = \frac{dp_{33}}{dt} = 0 \tag{8.33}$$

$$\frac{dp_{12}}{dt} = -\frac{p_{22}}{\tau_{12}}, \quad \frac{dp_{13}}{dt} = -\frac{p_{23}}{\tau_{12}}, \quad \frac{dp_{23}}{dt} = 0 \tag{8.34}$$

Suppose that at time $t = 0$ all the off-diagonal terms are zero. It is readily seen that, with $dp_{22}/dt = 0$, p_{12} becomes

$$p_{12}(t) = -p_{22}\frac{t}{\tau_{12}} \qquad (8.35)$$

and

$$p_{11}(t) = p_{11}(0) + p_{22}\left(\frac{t}{\tau_{12}}\right)^2 \qquad (8.36)$$

Similarly, with $dp_{23}/dt = 0$, it follows that

$$p_{13}(t) = 0$$

The skewness $p_{12}(t)$ of the thermal velocity w_1 grows linearly with time as particles arrive, with velocity w_2, from farther across the shear. Hence, p_{11} grows quadratically with time, representing the "viscous" heating as particles arriving from increasingly distant origins are transported across the shear by w_2. The presence of interparticle collisions would limit this growth, of course, and provide the familiar form of the viscosity. And, of course, the heat flow, omitted in this example of uniform shear, may produce further modification of the thermal motions.

In summary, the behavior of the six independent components of p_{ij} is easily understood in terms of the free motions of the individual particles across the bulk velocity gradients $\partial v_i/\partial x_j$. The same concepts come into play in chapter 9 where the Chew-Goldberger-Low calculation of p_{ij} is taken up in the presence of a large-scale magnetic field.

8.6 Effects of Collisions

Consider how the equations for p_{ij} are modified by scattering by interparticle collisions, by small-scale plasma waves, etc. A useful representation of the effects of scattering can be accomplished with a simple linear isotropic scattering term, so that in place of eqn. (8.20) we write

$$\frac{dp_{11}}{dt} = -p_{11}\left(\frac{3}{\tau_1} + \frac{1}{\tau_2} + \frac{1}{\tau_3}\right) + \frac{p_{22} - p_{11}}{\tau} + \frac{p_{33} - p_{11}}{\tau} \qquad (8.38)$$

$$\frac{dp_{22}}{dt} = -p_{22}\left(\frac{1}{\tau_1} + \frac{3}{\tau_2} + \frac{1}{\tau_3}\right) + \frac{p_{33} - p_{22}}{\tau} + \frac{p_{11} - p_{22}}{\tau} \qquad (8.39)$$

$$\frac{dp_{33}}{dt} = -p_{33}\left(\frac{1}{\tau_1} + \frac{1}{\tau_2} + \frac{3}{\tau_3}\right) + \frac{p_{11} - p_{33}}{\tau} + \frac{p_{22} - p_{33}}{\tau} \tag{8.40}$$

where τ is the characteristic scattering time, tailored to represent the particular conditions at hand. Note that in some special cases of wave scattering there may be different scattering times between different components, easily incorporated into the above equations instead of the single scattering time τ. An example is to be found in the solar wind, where the ion temperature is enhanced by wave heating (scattering) principally in the two transverse directions. This is evident from the fact that most of the expansion throughout interplanetary space is in the two transverse directions, yet it is observed that the transverse kinetic temperature is often as large or larger than the radial temperature. A representation of such net heat input is easily introduced into eqns. (8.38)–(8.40).

In other cases the anisotropy may drive plasma instabilities (waves) that scatter the particles so as to reduce the anisotropy. If left to itself, then, one expects no strong anisotropies to survive over the long characteristic times of the large-scale bulk motions. In the general case we go back to eqn. (8.15) and introduce into the right-hand side whatever linear scattering terms are appropriate for the problem at hand.

For the present conversation consider first the simple illustrative case that there is no overall compression or expansion ($1/\tau_1 = 1/\tau_2 = 1/\tau_3 = 0$), but the initial thermal velocity distribution is anisotropic. The subsequent relaxation to isotropy, described by eqns. (8.38)–(8.40), takes the form

$$p_{11}(t) = P + C_1 \exp\left(-\frac{3t}{\tau}\right) \tag{8.41}$$

$$p_{22}(t) = P + C_2 \exp\left(-\frac{3t}{\tau}\right) \tag{8.42}$$

$$p_{33}(t) = P + C_3 \exp\left(-\frac{3t}{\tau}\right) \tag{8.43}$$

where $C_1 + C_2 + C_3 = 0$, and P is given by

$$P = \frac{p_{11}(0) + p_{22}(0) + p_{33}(0)}{3} \tag{8.44}$$

There is a great variety of solutions to eqns.(8.38)–(8.40) for diverse combinations of τ_1, τ_2, τ_3, and τ, along with different initial conditions.

Indeed, in the presence of severe dilatation the relaxation time τ may be presumed to vary with the changing density N and effective kinetic temperature. Then, of course, the dilatation rates $1/\tau_n$ may vary with time. The essential point is that the present simple heuristic formulation of the dynamical equations for p_{ij} is about all one can hope to achieve when it comes to the computation of complex large-scale bulk motion of a plasma. Complete kinetic equations can be constructed by formal mathematical procedures, but they are too complicated to be of much use in dealing with realistic physical situations.

For many theoretical purposes, e.g., where the internal wave scattering is not known, one ignores the anisotropies and works out the gross features of the dynamics by treating the pressure as a simple scalar p, so that $p_{ij} = \delta_{ij}p$. The sum of eqns. (8.38)–(8.40) becomes

$$3\frac{dp}{dt} = -5p\left(\frac{1}{\tau_1} + \frac{1}{\tau_2} + \frac{1}{\tau_3}\right) \tag{8.45}$$

or

$$\frac{dp}{dt} + \gamma p\frac{\partial v_k}{\partial x_k} = 0 \tag{8.46}$$

where γ is the ratio of specific heats, equal to $\frac{5}{3}$ for particles with no participating internal degrees of freedom.

Heat flow is easily included by noting that the thermal energy density is $U = p/(\gamma - 1)$, so that in the absence of heat flow, eqn. (8.46) becomes

$$\frac{dU}{dt} + \gamma U\frac{\partial v_k}{\partial x_k} = 0 \tag{8.47}$$

In a collision-dominated plasma the thermal motions approach isotropy and there is a well-defined local temperature T. The heat flow can then be approximated by the usual thermal conduction term, writing

$$\frac{dU}{dt} + \gamma U\frac{\partial v_k}{\partial x_k} = \frac{\partial}{\partial x_k}\left(K_{ij}\frac{\partial T}{\partial x_j}\right) \tag{8.48}$$

The thermal conductivity may be anisotropic because of the presence of a magnetic field, so we write it here as a tensor K_{ij} . The thermal conductivity is strongly channeled along the magnetic field when the particle collision time exceeds the electron cyclotron frequency, because the conductivity is principally the work of the thermal electrons with their enormous thermal velocity (10^4 km/s at $T = 10^6$ K). Parallel to a magnetic field, or in all directions in the absence of a magnetic field, the thermal conductivity in the presence of collisons is

$$K = 6 \times 10^{-7} T^{5/2} \text{ ergs/cm s K} \tag{8.49}$$

for small thermal fluxes in ionized hydrogen (Chapman 1954; Spitzer 1956) with $K_{ij} = \delta_{ij} K$. The assumption is only that the mean free path is smaller than the scale of the magnetic field and the temperature field.

There are other sources of heat that might be included on the right-hand side of eqn. (8.48). In dense plasmas, radiative transfer of heat may be important, and can be approximated with a diffusion term similar to the term for thermal conductivity. For a detailed theory the reader is referred to Mihalas and Weibel-Mihalas (1984). Viscosity transfers kinetic energy of the bulk motion v into thermal motion w, although the effect is not large in most subsonic flows and can often be ignored. For a general discussion of the hydrodynamic energy equation the reader is referred to Landau and Lifschitz (1959).

In a collisionless plasma one has to sum over the trajectories of the individual ions and electrons, and the transfer of thermal energy becomes an exercise in Newtonian mechanics.

8.7 Off-diagonal Terms and Viscosity

The eqns. (8.35) and (8.36), for the time rate of change of the off-diagonal terms in the presence of uniform shear, are readily adapted to particle collisions by introducing the same linear scattering terms as in eqns. (8.38)–(8.40). With $\partial v_1/\partial x_2 = 1/\tau_{12}$ again, write

$$\frac{dp_{12}}{dt} = -\frac{p_{22}}{\tau_{12}} - \frac{p_{12}}{\tau}$$

$$\frac{dp_{13}}{dt} = -\frac{p_{23}}{\tau_{12}} - \frac{p_{13}}{\tau}$$

$$\frac{dp_{23}}{dt} = 0$$

where again τ is the characteristic particle scattering time. Putting $p_{23} = 0$, it follows that p_{13} declines in proportion to $\exp(-t/\tau)$ and is soon negligible if not initially zero. There remains p_{12}, and if p_{22} and τ_{12} are independent of time, we have

$$p_{12}(t) = -p_{22}\left(\frac{\tau}{\tau_{12}}\right)\left[1 - \exp\left(-\frac{t}{\tau}\right)\right]$$

for the initial condition that $p_{12}(0) = 0$. Thus, for $t >> \tau$,

$$p_{12}(t) \approx -p_{22}\frac{\tau}{\tau_{12}}$$

The $i = 1$ component of the momentum equation (8.12) becomes

$$NM\frac{\partial v_1}{\partial t} = -\frac{\partial p_{12}}{\partial x_2}$$

$$= +\frac{\partial}{\partial x_2}\left(p_{22}\tau\frac{\partial v_1}{\partial x_2}\right)$$

Thus, $p_{22}\tau$ represents the viscosity μ, and the kinematic viscosity is

$$\nu = \frac{p_{22}\tau}{NM}$$

Since $p_{22} = NM\langle w_2^2\rangle$ in terms of the mean square thermal velocity in the $i = 2$ direction, it follows that

$$\nu = \langle w_2^2\rangle\tau$$

as expected. The effective mean free path λ is $\langle w_2^2\rangle^{1/2}\,\tau$, so

$$\nu = \langle w_2^2\rangle^{1/2}\lambda \tag{8.50}$$

We can, of course, include the heating of p_{22} by the viscous dissipation, treating the scattering between p_{11}, p_{22}, and p_{33}. We then have a complete system of linear equations for dp_{ij}/dt.

8.8 Summary

This is as far as the conversation needs to go to outline the basis for the dynamical theory of the large-scale bulk motion of plasmas in preparation for understanding the magnetohydrodynamics of the cosmos. The development of the theory was based on the basic dynamical concepts of the bulk flow of a gas, with or without interparticle collisions, to emphasize the general and inescapable nature of the hydrodynamic equations for conservation of matter, conservation of momentum, and conservation of energy. These conditions provide a closed system of equations for N, v_i, p_{ij} through various prescriptions for computing the plasma pressure p or p_{ij}. More will be said on this later when we consider the effect of the magnetic field. The next step now is to introduce a large-scale magnetic field, with an eye to the simplest proper treatment of the physics with only the minimum essential mathematics.

9 Magnetohydrodynamics

9.1 Basic Considerations

As far as anyone is aware, there is no place in the cosmos where there is not a magnetic field, and there are only a few places, e.g., a cold planetary atmosphere or the deep interior of an exceedingly dense cold interstellar cloud, that lack free electrons and ions. So almost everywhere there is a magnetic field embedded in a plasma that freely conducts electric current. We noted in chapter 1 the remarkable ability of the agile electrons to transport electric charge, which is to say that the plasma generally has no significant electrical insulating properties and cannot support any significant electric field $\mathbf{E}'$ in its own moving frame of reference. This simple fact establishes magnetohydrodynamics as the proper description of the dynamics of the magnetic field. For with negligible $\mathbf{E}'$, eqn. (7.1) reduces to

$$\mathbf{E} = -\frac{\mathbf{v} \times \mathbf{B}}{c} + \mathbf{E}' \tag{9.1}$$

where $\mathbf{v}$ is the local bulk velocity of the plasma. Substituting this into Faraday's induction equation (6.3) leads to

$$\frac{\partial \mathbf{B}}{\partial t} = \nabla \times (\mathbf{v} \times \mathbf{B}) - c\nabla \times \mathbf{E}' \tag{9.2}$$

Then let $\mathbf{E}'$ go to zero, providing the familiar magnetohydrodynamic induction equation for an ideal fluid, which states that the magnetic field is carried bodily with the moving plasma. Having introduced the condition that $\mathbf{E}'$ is negligible, it must be appreciated that the resulting MHD equations, including the momentum equation, give dynamical results that are consistent with the initial assumption that $\mathbf{E}' = 0$.

Note again that there is an electric field $\mathbf{E} = -\mathbf{v} \times \mathbf{B}/c$ in the laboratory frame of reference *because* there is no electric field in the local frame of reference moving with the plasma. The laboratory field $\mathbf{E}$ is weak $O(v/c)$ compared to the magnetic field $\mathbf{B}$, and the stress and energy of the electric field are small $O(v^2/c^2)$ compared to the stress and energy of the magnetic field. Hence, again, the dynamical effects of the electric field are small to the same order as time dilatation and Lorentz-Fitzgerald contraction, both of which we neglect in nonrelativistic dynamics. It follows that the dynamical interplay of the plasma and magnetic field is a direct mechanical interaction between the magnetic stress, described by eqn. (4.10) or (7.12), and the inertia and pressure of the plasma, described in chapter 8. The dynamics of the large-scale magnetic field in the cosmos is an exercise in fluid mechanics combined with the induction equation (9.2) that states that the plasma and field move together.

Returning to eqn. (7.1) it follows that the magnetic field $\mathbf{B}'$ in the moving plasma is identical with the magnetic field $\mathbf{B}$ in the laboratory frame in the nonrelativistic approximation employed here, neglecting terms $O(v^2/c^2)$ compared to one.

The conventional proof that the magnetic field moves bodily with the plasma considers the time rate of change of the magnetic flux through a closed contour C' floating along with the bulk motion of the ideal plasma ($\mathbf{E}' = 0$). The calculation begins with the fixed closed contour C, with which C' is initially coincident. The time rate of change of the magnetic flux through C is given by

$$\begin{aligned}\frac{\partial}{\partial t}\int_C d\mathbf{S} \cdot \mathbf{B} &= \int_C d\mathbf{S} \cdot \frac{\partial \mathbf{B}}{\partial t} \\ &= \int_C d\mathbf{S} \cdot \nabla \times (\mathbf{v} \times \mathbf{B}) \\ &= \oint d\mathbf{s} \cdot \mathbf{v} \times \mathbf{B} \\ &= \oint \mathbf{B} \cdot d\mathbf{s} \times \mathbf{v}\end{aligned}$$

upon using the induction equation (9.2) and Stokes's theorem. Thus, in a time Δt the change in the magnetic flux through C is

$$\begin{aligned}\Delta\Phi &= \Delta t \frac{\partial}{\partial t}\int_C d\mathbf{S} \cdot \mathbf{B} \\ &= \oint \mathbf{B} \cdot d\mathbf{s} \times \mathbf{v}\Delta t \qquad (9.3)\end{aligned}$$

Figure 9.1 sketches the geometry of C and C'. It is evident that $\mathbf{v}\Delta t$ represents the displacement of the floating contour C' in the time Δt. With $d\mathbf{s}$ representing an element of length along C and C', the vector product $d\mathbf{s} \times \mathbf{v}\Delta t$ represents the area swept out by $d\mathbf{s}$ moving with C' in the time Δt. This states, then, that the change in the magnetic flux through C is

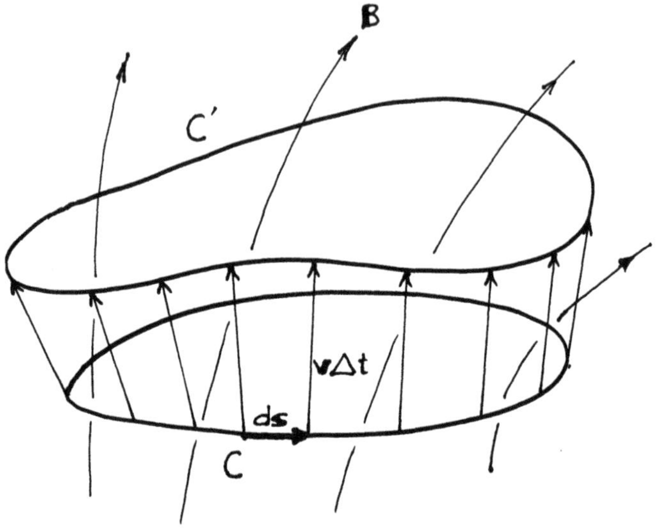

Fig. 9.1 A sketch of the lateral surface swept out by the contour C' floating freely with the fluid from its initial position C.

diverted out through the area swept out by C'. Since magnetic flux is conserved, i.e., $\nabla \cdot \mathbf{B} = 0$, it follows that there is no change of magnetic flux through the contour C' moving with the velocity $\mathbf{v}$ of the plasma. This follows for every contour C and its moving companion C', so it must be that the magnetic field moves bodily with the plasma. For if there were somewhere that the field failed to move with the plasma, the result (9.3) would be violated for a contour circling that area.

This is vividly illustrated upon using the special magnetic field profile sketched in Fig. 9.2. Thin sheets of magnetic flux are removed from the field so as to create markers in the field. Each moving thin layer of field-free plasma remains free of field as it is carried along with the plasma velocity $\mathbf{v}$. The regions of magnetic field between the moving field-free layers are defined by the field-free layers. So the plasma moves, and the field-free fiducial layers move with the plasma, and the field between the field-free layers remains between the field-free layers as the plasma moves. This is the essence of MHD, and it is the direct consequence of the electric field $\mathbf{E}'$ vanishing in the local moving frame of reference of the plasma.

As already noted, the Poynting vector corroborates the transport of the magnetic field with the bulk velocity $\mathbf{v}$ of the plasma. It follows from eqn. (6.4) that the electric field given by eqn. (9.1) represents the electromagnetic energy flux density

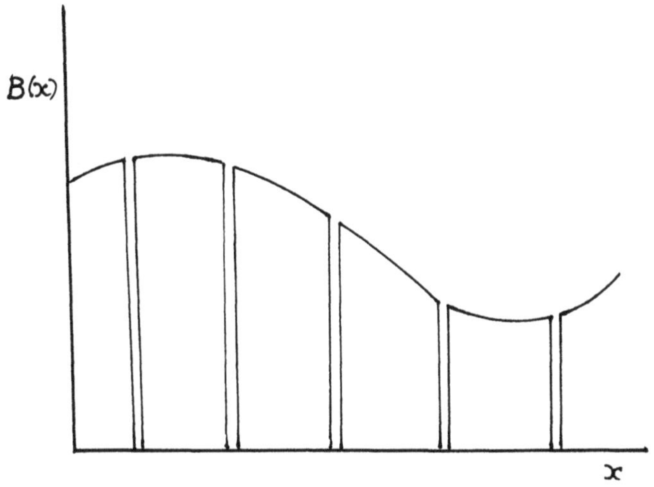

Fig. 9.2 A schematic drawing of the magnetic field intensity $B(x)$ from which thin sheets of flux have been removed at intervals to act as markers in the field.

$$\mathbf{P} = -\frac{(\mathbf{v} \times \mathbf{B}) \times \mathbf{B}}{4\pi} = \frac{vB^2 - \mathbf{B}\mathbf{v}\cdot\mathbf{B}}{4\pi} = \frac{\mathbf{v}_\perp B^2}{4\pi} \tag{9.4}$$

representing the bodily transport of the magnetic enthalpy $B^2/4\pi$ with the plasma velocity $\mathbf{v}_\perp$ perpendicular to **B**. To state this the other way around, the absence of electric field $\mathbf{E}'$ in the frame of reference of the plasma means that the Poynting vector $\mathbf{P}'$ also vanishes in the frame of reference of the plasma; i.e., there is no transport of electromagnetic energy relative to the plasma. The magnetic field is immovable in that frame of the plasma. Other demonstrations of the bulk transport of the magnetic field with the plasma can be constructed (cf. Parker 1979, pp. 32–42).

The magnetic field is obliged to move with the plasma, so the plasma experiences the magnetic stresses, described by the stress tensor M_{ij} given by eqn. (3.22) or (6.11). The momentum equation (8.12) for the plasma becomes

$$NM\left(\frac{\partial v_i}{\partial t} + v_j\frac{\partial v_i}{\partial x_j}\right) = -\frac{\partial p_{ij}}{\partial x_j} + \frac{\partial M_{ij}}{\partial x_j} \tag{9.5}$$

which is to be treated together with the induction equation (9.2) for the associated variation of B_i. The individual physical situation in the large-scale plasma in the cosmos is defined by the boundary conditions on B_i, v_i, and p_{ij}. Except in the ideal case of an infinitely conducting plasma, the boundary condition on B_i is continuity of both the normal and tangential components. In the ideal case, the tangential component may be discontinuous. The usual hydrodynamic considerations apply to v_i and to p_{ij}, remembering that the magnetic pressure $B^2/8\pi$ must be included along with p_{ij}. There is no need to be concerned with the boundary conditions of **E** and **j**, because they are automatically satisfied when the correct boundary conditions on **B** and **v** are applied.

Now Ampere's law, eqn. (3.9), tells us that there are electric currents associated with the deformation of the magnetic field. That means that in the presence of the small, but nonvanishing, electrical resistivity of the plasma there is a small electric field $\mathbf{E}'$ in the moving frame of the plasma. In chapter 1, eqn. (1.9), it was emphasized how small such electric fields are. The stress in $\mathbf{E}'$ is small $O(E'/B)$ compared to the magnetic stresses and negligible unless one treats a relativistic shock transition, or some similarly extreme phenomenon. The stress of the electric field **E** in the laboratory frame of reference is much stronger than the stresses in $\mathbf{E}'$, but still small $O(v^2/c^2)$ compared to the magnetic stresses. The stresses in the electric current **j** represent nothing more than the inertia of the conduction electrons, so **j**, **E**, $\mathbf{E}'$ play no role in the dynamics and are computed after the fact, once the dynamical equations are solved for **v** and **B**.

Note that in the confines of the plasma physics laboratory **E** may play a direct dynamical role. For instance, the experimenter may arbitrarily apply a potential difference of 10^5 V (3×10^2 statvolts) or more across 10 cm of plasma for the purpose of accelerating the plasma to the electric drift velocity $c\mathbf{E} \times \mathbf{B}/B^2$ (see appendix C), or to provide fast ions, or whatever. It has to be appreciated that there are generally no such interesting possibilities in astronomical settings where we have only the hydrodynamics of the plasma, driven by the magnetic stresses, the plasma pressure and inertia, and perhaps gravitational forces. Rapidly spinning neutron stars appear to be an important exception, wherein relativistic speeds in tenuous gases go far beyond the MHD of the lumbering plasma visible in the astronomical telescope.

9.2 Diffusion and Dissipation

The ideal induction equation (9.2) was constructed for the extreme case that the electric field $\mathbf{E}'$ in the moving plasma is identically zero. As already noted, $\mathbf{E}'$ is extremely small, but not really identically zero. The nonvanishing $\mathbf{E}'$ is responsible for the resistive dissipation of the magnetic field, neglected in eqn. (9.2). In most astronomical settings the dissipation is a very small effect, but sometimes it plays an essential role, as in rapid reconnection of magnetic field and in the MHD dynamo process in planets, stars, and galaxies. So we note again that, in a plasma sufficiently dense that the scalar Ohm's law applies, eqn. (1.7) provides the electric field

$$\begin{aligned} \mathbf{E}' &= \frac{\mathbf{j}}{\sigma} \\ &= \frac{c}{4\pi\sigma}\nabla \times \mathbf{B} \end{aligned}$$

where σ is the electrical conductivity, of course. Define the resistive diffusion coefficient $\eta \equiv c^2/4\pi\sigma$, so that eqn.(7.1) can be written

$$\mathbf{E} = -\frac{\mathbf{v} \times \mathbf{B}}{c} + \frac{\eta}{c}\nabla \times \mathbf{B}$$

The induction eqn. (9.2) becomes

$$\frac{\partial \mathbf{B}}{\partial \mathbf{t}} = \nabla \times (\mathbf{v} \times \mathbf{B}) - \nabla \times (\eta\nabla \times B) \tag{9.6}$$

$$= \nabla \times (\mathbf{v} \times \mathbf{B}) + \eta\nabla^2\mathbf{B} + \mathbf{B} \times \nabla\eta \tag{9.7}$$

Equation (9.7) is customarily simplified by ignoring the effects of $\nabla\eta$, so that the dissipation term takes the form of the Laplacian. The effect of resistivity is, then, simple diffusion, with the familiar form

$$\frac{\partial \mathbf{B}}{\partial \mathbf{t}} = \nabla \times (\mathbf{v} \times \mathbf{B}) + \eta\nabla^2\mathbf{B} \tag{9.8}$$

for the induction equation. In a motionless plasma the individual Cartesian components of the magnetic field independently diffuse according to

$$\frac{\partial B_i}{\partial t} = \eta\nabla^2 B_i \tag{9.9}$$

With $\sigma = 2 \times 10^7\ T^{3/2}$/s, it follows that

$$\eta = \frac{4 \times 10^{12}}{T^{3/2}}\ \mathrm{cm}^2/\mathrm{s} \tag{9.10}$$

for ionized hydrogen. Thus, $\eta = 4 \times 10^6$ cm²/s at 10^4 K, and 4×10^3cm²/s at coronal temperatures of 10^6 K. Diffusion over a scale L takes place in a characteristic time $t = L^2/4\eta$, so over a scale of 100 km (10^7cm) in the corona, diffusion alone takes place in about 6×10^9 s, or 2×10^2 years. In fact, observations show that small-scale magnetic configurations sometimes change dramatically in an hour or so, demonstrating the importance of the plasma motions and the phenomenon of dynamical rapid reconnection. We will have more to say on this in chapter 10.

So, as we noted earlier, the electric field $\mathbf{E}'$ experienced by the plasma is responsible for the dissipation and diffusion of the magnetic field while playing no direct active role in the dynamics.

9.3 Application of Magnetic Diffusion

The diffusion term on the right-hand side of eqn. (9.8) is small compared to the term $\nabla \times (\mathbf{v} \times \mathbf{B})$ representing the convection of the magnetic field with the plasma velocity $\mathbf{v}$ in the large-scale plasma phenomena of astrophysics, which includes the dynamics of planetary magnetospheres. For fields with characteristic scale l the convection term is $O(vB/l)$, while the diffusion term is $O(\eta B/l^2)$. The ratio of convection to diffusion is the well-known magnetic Reynolds number,

$$R_{\mathrm{m}} = \frac{vl}{\eta} \tag{9.11}$$

Choosing the characteristic velocity v to be the Alfvèn speed $C = B/(4\pi\rho)^{1/2}$ provides the Lundquist number,

$$R_{\mathrm{l}} = \frac{Cl}{\eta} \tag{9.12}$$

The Lundquist number compares the characteristic diffusion time l^2/η across the scale l directly with the dynamical Alfvèn crossing time l/C. Both the magnetic Reynolds number and the Lundquist number are large compared to one in the large-scale MHD activity in the cosmos; i.e., the overall diffusion is very small compared to the convection of the magnetic fields.

It must be appreciated, however, that the special exceptions to this general fact play an important role. The outstanding exception arises in the phenomenon of rapid magnetic reconnection, wherein the magnetic stresses expel plasma from between two oppositely directed components of the magnetic field, thereby pushing the dominant scale l asymptotically toward zero. Diffusion and dissipation become large in the thin current sheet, of thickness l, between the two regions of opposite field, eating into the field on either side at speeds up to some fraction of the Alfvèn speed C. The magnetic flare phenomenon is the extreme example of rapid reconnection. That important topic is taken up in section 10.6.

Another important exception is the photosphere of the Sun, where the degree of ionization falls so low that η may be as large as 10^{10}cm^2/s (σ = 10 ^{10}/s) (Kopecky and Obridko 1968; Kopecky and Kuklin 1969) in a layer with characteristic thickness of the order of 10^7cm. The characteristic diffusion time $l^2/4\eta$ is about an hour over the diameter $l = 10^7$cm of one of the photospheric magnetic fibrils. The characteristic Alfvèn speed is 10^6cm/s in the fibril field of 1500 G and a gas density of the order of 2×10^{-7}g/cm^3. The Lundquist number is only about 250. Thus, neighboring fibrils may merge slowly, but the overall integrity of the magnetic flux is preserved by the much smaller resistivity of the hotter layers immediately above and below the stratum of relatively high resistivity. On the other hand, the rapid reconnection between two oppositely directed fibrils may produce a microflare. Equally important is the steady inflow of gas in the resistive stratum, with a characteristic speed of the order of 4η divided by the fibril radius, or close to 10^4cm/s. Presumably this inflow produces a cascade of gas downward within the fibril.

In contrast to these exceptions, the Lundquist number for the large-scale magnetic fields of a star like the Sun is so large as to avoid any significant dissipation apart from the phenomenon of rapid reconnection. Thus, for instance, the mean 10^2 G field of a bipolar magnetic region on the Sun may have a characteristic scale of 10^{10}cm. At the chromospheric level, where T is of the order of 10^4 K and the density NM is perhaps 10^{-12} g/cm^3, the Alfvèn speed C is 2×10^7cm/s and the Lundquist number R_L is 0.5×10^{11}. In the corona, where T is 10^6 K and NM is equal to 2×10^{-14} g/cm^3, the Alfvèn speed is 2×10^8cm/s and R_L is about 0.5×10^{15}. Even the thinnest detectable striations, with l as small as 10^7cm, have Lundquist numbers of 0.5×10^8 in the chromosphere and 0.5×10^{12} in the corona. It follows that the observed rapid evolution of the active coronal bipolar fields involves both convective deformation and rapid reconnection, because overall diffusion alone requires a time $l^2/4\eta = 6 \times 10^9$s, or about 200 years, in the thin striations, as already noted, and vastly longer in the field at large.

Having remarked, then, on the minor role of resistive diffusion in the large-scale magnetic fields in the astronomical universe, we must not overlook the fact that diffusion is an essential part of the $\alpha\omega$-dynamo operation, believed to be the origin of the magnetic field of Earth, the magnetic field of the Sun, and, evidently, even the magnetic field of the

Galaxy. A magnetic Reynolds number of the order of 10^2–10^3 is about right for the efficient operation of the dynamo in its basic mode. Thus, for instance, in the liquid iron core of Earth, where $\sigma \approx 10^{16}$/s and $\eta \approx 10^4$cm^2/s, the fluid motions are of the order of 0.03 cm/s on scales of $l = 10^8$ cm. The magnetic Reynolds number turns out to be 300. In effect, the field diffuses across the scale l in about 10^4 years while the nonuniform rotation of the fluid core moves one revolution around the core (radius 3×10^8 cm) in roughly the same period of time.

However, a serious problem arises with the same $\alpha\omega$-dynamo process that is presumed to produce the periodic magnetic fields of the Sun. Detailed numerical dynamo models readily duplicate the solar magnetic cycle inferred from the 11- and 22-year variation of the magnetic fields observed at the surface of the Sun. An essential part of that duplication is the assumption of an effective diffusion coefficient η of the order of 10^{11}cm^2/s. Basically, the field has to diffuse a distance comparable to half the depth of the convective zone, or the width of a band of toroidal magnetic field, amounting to $l = 1 \times 10^{10}$cm, in the 11-year solar half-cycle. This requires a diffusion coefficient in the vicinity of 10^{11}cm^2/s. However, deep in the convective zone, where the dynamo is presumed to operate, T is of the order of 10^6 K and the resistive diffusion coefficient η is only 4×10^3cm^2/s. Even a few hundred kilometers below the visible surface, where T is as small as 10^4 K, the diffusion coefficient is only 4×10^6cm^2/s. So there is a qualitative problem here. The customary rationalization is that turbulent diffusion of the magnetic fields dominates the ordinary resistive diffusion. The mixing length representation of turbulence, i.e., dimensional analysis of the turbulent convection beneath the surface of the Sun, suggests an effective turbulent diffusion coefficient $\eta_t \approx 0.1lv$ for eddies with scale l and characteristic velocity v. Thus, for instance, the granules at the visible surface exhibit scales of 500 km and characteristic velocities of the order of 0.5 km/s, yielding an estimated diffusion coefficient $\eta_t \approx 2.5 \times 10^{11}$cm^2/s. Deeper in the convective zone the scale l increases while the velocity decreases, and the resulting magnetic Reynolds number remains close to 10^{11}cm/s, diminishing significantly only as the bottom of the convective zone is approached. So it would seem plausible that the magnetic field of the Sun is generated and reformed by turbulent dissipation every 11 years.

Unfortunately, the idea of turbulent diffusion becomes difficult to comprehend (Parker 2001) when it is realized that the observational inventory of magnetic flux emerging in a long-lived activity complex (Gaizauskas et al. 1983) over a year or so involves a total in excess of 10^{23} maxwells. Unless the Sun has some clever way of recycling magnetic flux, this implies that there are more than 10^{23} maxwells in the azimuthal magnetic field in the lower convective zone, where it might reasonably be stored. Over a meridional cross-sectional area of 5×10^9cm by 1×10^{10}cm (about 20° of latitude) this requires a mean field not less than 2×10^3 G. This is very nearly as strong as the maximum equipartition field of about 3×10^3 G, at which the Maxwell stresses of the field are as great as the Reynolds stresses of the convective turbulence. So how can the convection freely mix and diffuse the mean large-scale field? Any swirling of the field greatly increases the field strength, which is already nearly as strong as the turbulence.

The same dilemma exists with the galactic dynamo, where a diffusion coefficient of the order 10^{25}cm^2/s is needed for the theory of the galactic dynamo. Mixing length theory suggests a turbulent diffusion coefficient of about 0.3×10^{25}cm^2/s, based on a characteristic scale $l = 10$ pc and a characteristic interstellar turbulent velocity $v = 10$ km/s. The galactic magnetic field is of the order of 4×10^{-6} G, again comparable to the equipartition field of the interstellar turbulence. So we have no idea how significant turbulent mixing can take place. We have conjectured that the cosmic rays play a dynamical role, inflating the unstable galactic field and extending lobes of field far out from the gaseous disk to form the galactic halo. Rapid reconnection between adjacent bipolar lobes may provide the necessary dissipation and escape of magnetic flux (Parker 1992). But this is entirely speculative.

The essential point is that we have some distance to go before we understand the $\alpha\omega$ dynamo in the context of stars and galaxies, and, until we fully understand the dynamo effect, how can we apply it to the many unresolved, but magnetically active, objects observed in the cosmos?

9.4 Discussion

To take stock of where we are in this conversation, it is clear that the magnetic and electric fields in the cosmos have a remarkable form. There is the uniquely defined magnetic field, with $\mathbf{B}' = \mathbf{B}$ in every moving frame of reference ($v \ll c$), while there are different electric fields in each of the different moving reference frames. The electric field is essentially zero in the swirling nonuniform frame of the moving plasma, which defines the electric field to be $\mathbf{E} = -\mathbf{v} \times \mathbf{B}/c$ at any given point in any reference frame with velocity $-\mathbf{v}$ relative to the plasma. So the choice of the laboratory reference frame determines the laboratory electric field **E**.

As already noted, the dynamics is described by eqns. (9.2) and (9.5), so the outcome is determined by the interaction of the inertia of the plasma velocity **v**, the plasma pressure p_{ij}, and the Maxwell stress M_{ij} in the magnetic field. The initial conditions, together with the boundary conditions on **v** and **B** provide a unique dynamical state for the system.

Now one sometimes comes across the view that the electric current **j** and electric field **E** must also be considered with their own special boundary conditions (cf. Melrose 1995; Parks 2004). This arbitrary view overlooks the fact that the boundary conditions on **v** and **B** have already determined a unique solution, so there is no remaining freedom to incorporate any further independent boundary conditions (Parker 1996b). In any case the proponents will find that **j** and **E** are readily computed from the solutions for **v** and **B**, and the boundary conditions on **j** and **E** are automatically satisfied by the solutions of eqns. (9.2) and (9.5) based on the boundary conditions on **v** and **B**. This built in self-consistency is a universal property of physics, for without it there would be contradictions between fundamental laws of physics, meaning that one or more of those laws was incorrectly stated.

With these remarks we turn the conversation to the diverse circumstances under which MHD applies in astronomical settings. First we have a look at the dynamics of a magnetic field in a gas that is only slightly ionized, showing that, if there are enough free electrons to provide a high electrical conductivity, then MHD is applicable to the large-scale dynamics. The Hall effect, the Pederson diffusion coefficient (ambipolar diffusion), and the ordinary resistivity are all small effects in large-scale magnetic fields. Following that exercise, we have a look at the ideal case of a collisionless plasma. There is no resistive dissipation in a collisionless plasma, and electric currents do not follow simply from an electric field **E**′ in the local frame of reference of the moving plasma. So some authors have been concerned that there is a problem in satisfying Ampere's law. One need not worry, of course, because Ampere's law and Maxwell's equations are both fundamental laws of nature. So they are never violated. However, it is interesting to see how the automatic conformity to Ampere and Maxwell actually comes about.

9.5 Partially Ionized Gases

Some of the gas in the cosmos is partially ionized and some is only very slightly ionized so there are effects in addition to the simple resistivity included in the MHD induction eqns. (9.5) and (9.8). The terrestrial ionosphere is a case in point, in which the degree of ionization increases from zero, at the top of the stratosphere, to essentially fully ionized at an altitude of 10^3 km. To have a look at the effects of a significant neutral component, we turn to a three-component gas, with the simplification that the gas is only very slightly ionized so that the neutral component represents the principal mass density.

Denote the number density of the neutral gas by N, the mass of a molecule by M, and the local mean bulk velocity of the neutral gas by **v**. The ions have each lost one electron, of mass m ($\ll M$), and the number density of both the ions and electrons is n ($\ll$ N) so as to maintain overall electrical neutrality. Let the mean bulk velocity of the ions be $\mathbf{w}$, with $\mathbf{u}$ for the electrons, not to be confused with the use $\mathbf{u} = \mathbf{v} + \mathbf{w}$ in chapter 8. The neutral gas and the electrons and ions are coupled by interparticle collisions, so the equation of motion for the neutral gas can be written

$$NM\frac{d\mathbf{v}}{dt} = -\nabla p + \frac{nM(\mathbf{w} - \mathbf{v})}{\tau_i} + \frac{nm(\mathbf{u} - \mathbf{v})}{\tau_e} + N\mathbf{F} \quad (9.13)$$

where p is the isotropic gas pressure, $\mathbf{F}$ is any external force, e.g., gravity, applied to the individual molecule. The time over which an ion makes a collision with a neutral atom is denoted by τ_i, while the time over which an electron collides with a neutral atom is denoted by τ_e. We are again using the simple linear friction representation of the collisional interaction between particle species.

The equation of motion for the ions can be written

$$nM\frac{d\mathbf{w}}{dt} = -\nabla p_i + ne\left(\mathbf{E} + \frac{\mathbf{w} \times \mathbf{B}}{c}\right) - \frac{nM(\mathbf{w} - \mathbf{v})}{\tau_i} - \frac{nm(\mathbf{w} - \mathbf{u})}{\tau} + n\mathbf{f}_i \quad (9.14)$$

where τ is the time over which an electron collides with an ion, and $\mathbf{f}_i$ is whatever external force is applied to the individual ion. The ion pressure is denoted by p_i, and assumed to be isotropic. The generalization to a tensor p_{ij} is obvious, if needed. The mean electric field E' experienced by the ion is $\mathbf{E} + \mathbf{w} \times \mathbf{B}/c$.

For the electrons

$$nm\frac{d\mathbf{u}}{dt} = -\nabla p_e - ne\left(\mathbf{E} + \frac{\mathbf{u} \times \mathbf{B}}{c}\right) - \frac{nm(\mathbf{u} - \mathbf{v})}{\tau_e} + \frac{nm(\mathbf{w} - \mathbf{u})}{\tau} + n\mathbf{f}_e \quad (9.15)$$

where p_e is the electron pressure, and the mean electric field $\mathbf{E}_e'$ experienced by the electrons is $\mathbf{E} + \mathbf{u} \times \mathbf{B}/c$. The external force exerted on the individual electron is denoted by $\mathbf{f}_e$.

Now in the simplest case, the inertia of the ions and electrons, as well as their pressures, are neglected. That is to say, the electrons and ions are assumed to be cold, and their motion is driven primarily by the Lorentz force while being tied to the relatively dense neutral gas by collisional friction. So in carrying through the algebra it is convenient to define the small quantities

$$nG = -\nabla p_i - nM\frac{d\mathbf{w}}{dt} + n\mathbf{f}_i \tag{9.16}$$

$$nH = -\nabla p_e - nm\frac{d\mathbf{u}}{dt} + n\mathbf{f}_e \tag{9.17}$$

for the subsequent algebraic manipulations.

We proceed by solving the ion and electron equations of motion and Ampere's law for **u** and **w** and the electric field **E**. Thus, **u** and **w** can be eliminated from eqn. (9.13) for the motion of the neutral gas. Substituting **E** into Faraday's induction equation provides the equation for $\partial\mathbf{B}/\partial t$.

To begin, note that the electric current density is $\mathbf{j} = ne(\mathbf{w} - \mathbf{u})$, so that Ampere's law, eqn. (3.11), becomes

$$\mathbf{u} = \mathbf{w} - \frac{c\nabla \times \mathbf{B}}{4\pi ne} \tag{9.18}$$

Then add the ion and electron equations of motion, with the result that

$$\frac{nM(\mathbf{w} - \mathbf{v})}{\tau_i} + \frac{nm(\mathbf{u} - \mathbf{v})}{\tau_e} = \frac{\mathbf{j} \times \mathbf{B}}{c} + nG + nH \tag{9.19}$$

With this expression for the collision terms, the equation of motion (9.13) for the neutral gas takes the obvious form

$$NM\frac{d\mathbf{v}}{dt} = -\nabla p + \frac{(\nabla \times \mathbf{B}) \times \mathbf{B}}{4\pi} + NF + nG + nH \tag{9.20}$$

upon using Ampere's law to replace **j** by $\nabla \times \mathbf{B}$. The correction to the applied force **F** is $n(G + H)$, as we would expect.

The next step is to solve eqns. (9.18) and (9.19) for **u** and **w**, with the result

$$\mathbf{w} = \mathbf{v} + \frac{cm/\tau_e}{4\pi neQ}\nabla \times \mathbf{B} + \frac{(\nabla \times \mathbf{B}) \times \mathbf{B}}{4\pi nQ} + \frac{G + H}{Q} \tag{9.21}$$

$$\mathbf{u} = \mathbf{v} - \frac{cM/\tau_i}{4\pi neQ}\nabla \times \mathbf{B} + \frac{(\nabla \times \mathbf{B}) \times \mathbf{B}}{4\pi nQ} + \frac{G + H}{Q} \tag{9.22}$$

where $Q = M/\tau_i + m/\tau_e$. Then solve the equation of motion for the ions (9.14), obtaining the electric field **E**,

$$\mathbf{E} = -\frac{\mathbf{G}}{e} - \frac{\mathbf{w} \times \mathbf{B}}{c} + \frac{\mathbf{w}}{e}\left(\frac{M}{\tau_i} - \frac{m}{\tau}\right) + \frac{m}{e\tau}\mathbf{u} - \frac{M}{e\tau_i}\mathbf{v}$$

Then use eqns. (9.21) and (9.22) to eliminate **w** and **u**, so that

$$\mathbf{E} = -\frac{\mathbf{v} \times \mathbf{B}}{c} + \frac{c}{4\pi ne^2}\left[\frac{m}{\tau} + \frac{(m/\tau_e)(M/\tau_i)}{Q}\right]\nabla \times \mathbf{B}$$
$$+ \frac{(M/\tau_i - m/\tau_e)}{4\pi neQ}(\nabla \times \mathbf{B}) \times \mathbf{B} - \frac{[(\nabla \times \mathbf{B}) \times \mathbf{B}] \times \mathbf{B}}{4\pi ncQ}$$
$$+ \frac{\mathbf{H}M/\tau_i}{eQ} - \frac{\mathbf{G}m/\tau_e}{eQ} - \frac{(\mathbf{H} + \mathbf{G}) \times \mathbf{B}}{cQ} \qquad (9.23)$$

The contributions of **G** and **H** are clearly displayed here in relation to the other terms, and their magnitudes can be estimated to see if they can be neglected. Their principal effects are those of gravity, and ion and electron pressures. To keep the equations from becoming inconveniently lengthy, we drop **G** and **H** at this point and proceed to consider only the basic Ohmic, Hall, and Pederson contributions. The reader can easily carry **G** and **H** along to the final result, if desired.

It is convenient to introduce the coefficients α, β, and η, with the Hall coefficient written

$$\alpha = cB\frac{M/\tau_i - m/\tau_e}{4\pi neQ} \qquad (9.24)$$

and the Pedersen coefficient

$$\beta = \frac{B^2}{4\pi nQ} \qquad (9.25)$$

The Pedersen resistivity is another term for ambipolar diffusion, resulting from the fact that the Lorentz force, $(\nabla \times \mathbf{B}) \times \mathbf{B}/4\pi$, is exerted only on the ions and electrons, thereby driving them through the neutral gas in opposition to the frictional drag, to be seen on the right-hand sides of eqns. (9.14) and (9.15) (cf. Parker 1979, pp. 45, 117). As a result, the magnetic field slips slowly through the neutral gas as it drives the ions and electrons along.

The Ohmic resistive diffusion coefficient is defined as

$$\eta = \frac{c^2}{4\pi n e^2}\left[\frac{m}{\tau} + \frac{(m/\tau_e)(M/\tau_i)}{Q}\right] \tag{9.26}$$

The quantity B introduced into the Pedersen coefficient represents a specified characteristic field strength, introduced to make a comparison of the Hall, Pedersen, and Ohmic contributions. Write the dimensionless magnetic field as $\mathbf{b} = \mathbf{B}/B$, so that

$$\begin{aligned}\mathbf{E} = \frac{B}{c}\{&-\mathbf{v} \times \mathbf{b} + \eta\nabla \times \mathbf{b} + \alpha(\nabla \times \mathbf{b}) \times \mathbf{b} \\ &- \beta[(\nabla \times \mathbf{b}) \times \mathbf{b}] \times \mathbf{b}\}\end{aligned} \tag{9.27}$$

Substituting this expression for $\mathbf{E}$ into the Faraday induction equation yields

$$\begin{aligned}\frac{\partial \mathbf{b}}{\partial t} = {}& \nabla \times (\mathbf{v} \times \mathbf{b}) - \nabla \times (\eta\nabla \times \mathbf{b}) - \nabla \times [\alpha(\nabla \times \mathbf{b}) \times \mathbf{b}] \\ &+ \nabla \times \{\beta[(\nabla \times \mathbf{b}) \times \mathbf{b}] \times \mathbf{b}\}\end{aligned} \tag{9.28}$$

It is convenient to introduce the dimensionless Lorentz force,

$$\mathbf{L} = \frac{(\nabla \times \mathbf{b}) \times \mathbf{b}}{4\pi} \tag{9.29}$$

to show its role in the Hall and Pedersen effects. The electric field, eqn. (9.27), becomes

$$\mathbf{E} = \frac{B}{c}(-\mathbf{v} \times \mathbf{b} + \eta\nabla \times \mathbf{b} + \alpha\mathbf{L} - \beta\mathbf{L} \times \mathbf{b}) \tag{9.30}$$

and the induction equation (9.28) becomes

$$\begin{aligned}\frac{\partial \mathbf{b}}{\partial t} = {}& \nabla \times (\mathbf{v} \times \mathbf{b}) - \nabla \times (\eta\nabla \times \mathbf{b}) - \nabla \times \alpha\mathbf{L} \\ &+ \nabla \times [\beta(\mathbf{L} \times \mathbf{b})]\end{aligned} \tag{9.31}$$

The Hall effect is a consequence of an electric field in the direction of $\mathbf{L}$ necessary to maintain electrical neutrality. It is evident from eqn. (9.24) that the coefficient α arises from the difference in the frictional drag of the neutral molecules on the ions and electrons, $M/\tau_i - m/\tau_e$. So there must be an electric field in the direction in which the ions and electrons are driven by **L** if they are to move together. The Pedersen effect (ambipolar diffusion) arises from the drift of the ions and electrons through the neutral molecules driven by the Lorentz force. The magnetic field does work on the ions and electrons as it pushes them with force **L** through the neutral gas, and the Pedersen effect represents the loss of magnetic energy as the field performs that work.

To see this directly from the equations, note that the magnetic energy equation follows from the scalar product of **b** with the induction equation,

$$\frac{\partial}{\partial t}\frac{b^2}{8\pi} = -\mathbf{L}\cdot\mathbf{v} - 4\pi\beta L^2 - \frac{\eta(\nabla\times\mathbf{b})^2}{4\pi} - \nabla\cdot(\eta\mathbf{L} + \beta b^2\,\mathbf{L} + \alpha\mathbf{L}\times\mathbf{b}) \tag{9.32}$$

The first term on the right-hand side represents the work done on the ions and electrons by the Lorentz force **L**. The second term represents the work done by the slippage of the ions and electrons through the neutral component. It is quadratic in **L** because the rate of slippage is proportional to L, so the rate of doing work is proportional to L^2. The third term is the familiar resistive dissipation. The fourth term is the divergence of the energy flux

$$\mathbf{W} = \eta\mathbf{L} + \beta b^2\mathbf{L} + \alpha\mathbf{L}\times\mathbf{b}$$

If $\nabla\cdot\mathbf{W}$ is integrated over a volume V so large that the magnetic field vanishes everywhere on the surface of V, then the surface integral of **W** vanishes and so does the volume integral of $\nabla\cdot\mathbf{W}$. It follows that **W** represents a circulation of energy within the volume V, with no net loss of energy. The Hall effect represents no net dissipation, contributing only to the circulation and redistribution of energy within the overall volume of field.

The essential point is that the dissipative effects of Ohmic and Pedersen diffusion, as well as the nondissipative Hall effect, involve two spatial derivatives, i.e., second order in ∇, or in the inverse large-scale $1/l$. The induction term $\nabla\times(\mathbf{v}\times\mathbf{b})$, on the other hand, is only first order in ∇. So in the dynamics of large-scale magnetic fields the dominant term is the induction term and we have the familiar MHD. Only in extreme cases are α, β, and η large enough, and the scale l sufficiently small, to contribute to the overall dynamics. Needless to say, in the presence of the small scales to be found across shock fronts, current sheets in magnetic reconnection, laboratory confinement experiments, etc., the effects of α, β, and η may be important.

9.6 An Electric Current to Satisfy Ampere

This is an appropriate place to consider the question that sometimes arises when working with the magnetic field **B** and bulk plasma velocity **v**, as to how one can be sure that Ampere's law, eqn. (3.2) or (3.11), is satisfied. To begin, note quite generally that if the electric current **j** is parallel to $\nabla \times \mathbf{B}$ but does not have the appropriate magnitude, so that it does not satisfy Ampere's law—say

$$4\pi \mathbf{j} = c\nabla \times \mathbf{B}(1 - \varepsilon)$$

where $\varepsilon \ll 1$—then Maxwell's equation (6.7) becomes

$$\frac{\partial \mathbf{E}}{\partial t} = \varepsilon c \nabla \times \mathbf{B}$$

stating that **E** increases rapidly with the passage of time in an effort to push more current in the direction of $\nabla \times \mathbf{B}$ so as to satisfy Ampere's law. The growth of **E** goes on without limit until Ampere's law is satisfied. So in collision-dominated plasmas, there is no problem. It was emphasized in section 1.3 that the electrical conductivity is generally sufficiently large that an extremely weak electric field $\mathbf{E}'$ suffices to provide a current in compliance with Ampere. A more interesting situation is the ideal collisionless plasma, where there is no Ohm's law, and each ion and electron follows its own inexorable trajectory, predetermined by Newton's equations of motion. What happens?

Parallel to the magnetic field the electric field easily accelerates the numerous ions and electrons to produce the necessary electric current, along the lines described in section 1.3. The dynamics in the two transverse directions is more complicated. The general motion of a charged particle, with mass m and charge q is displayed in appendix C for arbitrary $E(t)$ perpendicular to B. Here we need to look more closely at the net effect of the particle motions. So, with a slowly varying uniform magnetic field $\mathbf{B}(t)$ in the z direction and a slowly varying uniform electric field $\mathbf{E}(t)$ in the x direction, we have

$$\frac{d^2x}{dt^2} = \frac{e}{m}E(t) + \Omega(t)\frac{dy}{dt} \tag{9.33}$$

$$\frac{d^2y}{dt^2} = -\Omega(t)\frac{dx}{xt} \tag{9.34}$$

where the cyclotron frequency is given by $\Omega(t) = qB(t)/mc$. Divide eqn. (9.33) by $\Omega(t)$, obtaining

$$\frac{1}{\Omega(t)}\frac{d^2x}{dt^2} = c\frac{E(t)}{B(t)} + \frac{dy}{dt} \tag{9.35}$$

Average this equation over a cyclotron period taking advantage of the slow variation of the large-scale fields $E(t)$ and $B(t)$. The rapidly oscillating left-hand side vanishes, leaving

$$\left\langle\frac{dy}{dt}\right\rangle = -c\frac{E(t)}{B(t)} \tag{9.36}$$

This is the so called electric drift velocity $\mathbf{u}$, written generally as

$$\mathbf{u} = c\frac{\mathbf{E}\times\mathbf{B}}{B^2} \tag{9.37}$$

This mean drift velocity is charge independent, and both electrons and ions participate equally, so it represents the bulk motion of the plasma. The essential feature is that the motion $\mathbf{u}$ places the plasma in the local frame of reference in which there is no electric field, $\mathbf{E}' = 0$. Only in that frame of reference can the ions and electrons move together,

Note, then, that $\mathbf{u}$ represents the bulk velocity of the plasma driven by the external imposition of the electric field $\mathbf{E}(t)$ perpendicular to $\mathbf{B}(t)$. On the other hand, in astronomical settings, where there is no external interference, the bulk velocity $\mathbf{u}$ is determined by Newton's momentum equation (8.12). In that case eqn. (9.37) determines $\mathbf{E}$ in the laboratory frame of reference from the fact that $\mathbf{E}' = 0$ in the plasma. Solving eqn. (9.37) for $\mathbf{E}$ in the laboratory frame of reference yields the familiar relation

$$\mathbf{E}_\perp = -\frac{\mathbf{u}\times\mathbf{B}}{c} \tag{9.38}$$

for the component of the electric field perpendicular to $\mathbf{B}$.

Note, then, that the individual ion or electron, with thermal velocity $\mathbf{w}$ relative to the bulk motion $\mathbf{u}$, experiences the electric field $\mathbf{w}\times\mathbf{B}/c$, i.e. the Lorentz force, causing it to execute the familiar cyclotron motion.

Now suppose that we divide eqn. (9.33) by $\Omega(t)$. Then differentiate the result with respect to time, and use eqn. (9.34) to eliminate d^2y/dt^2, obtaining

$$\frac{d}{dt}\left[\frac{1}{\Omega(t)}\frac{d^2x}{dt^2}\right] = \frac{d}{dt}\left[c\frac{E(t)}{B(t)}\right] - \Omega(t)\frac{dx}{dt}$$

Averaging over the cyclotron period eliminates the rapidly oscillating left-hand side, so that

$$\left\langle \frac{dx}{dt} \right\rangle = \frac{d}{dt}\left[c\frac{E(t)}{B(t)} \right]\frac{1}{\Omega(t)}$$
$$= \frac{du}{dt}\frac{1}{\Omega(t)}$$

where u is the magnitude of the electric drift velocity $\mathbf{u}$, given by eqn. (9.37): The motion $<dx/dt>$ is referred to as the *polarization drift* and denoted by $\mathbf{u}_p$, with

$$\mathbf{u}_p = \frac{1}{\Omega}\frac{\mathbf{B}}{B} \times \frac{d\mathbf{u}}{dt} \tag{9.39}$$

It is evident that the mean particle drift $<dx/dt>$ depends on the sign of q in $\Omega(t)$, so the ions and electrons drift in opposite directions, producing a net current. This polarization drift arises from the inertial force exerted on each ion and electron by the bulk acceleration $d\mathbf{u}/dt$. The associated polarization electric current is

$$\mathbf{j}_p = \frac{n(M + m)c}{B^2}\mathbf{B} \times \frac{d\mathbf{u}}{dt} \tag{9.40}$$

where now M is the ion mass and m the electron mass. It is evident that the electric field drives an electric current for only so long as the plasma is accelerated. So it cannot go on indefinitely. On the other hand, $\nabla \times \mathbf{B}$ may well be nonvanishing in a direction perpendicular to $\mathbf{B}$. That is to say, so long as the plasma is pushing against the magnetic field, or vice versa, there is a bulk Lorentz force, with a component of $\nabla \times \mathbf{B}$ perpendicular to $\mathbf{B}$, requiring a flow of current perpendicular to $\mathbf{B}$. Evidently, it is necessary to look further into the motion of the individual ions or electrons, because the presence of a Lorentz force must somehow produce the associated current required by Ampere's law.

First of all, the electrons and ions all drift in the frame of reference in which there is no electric field—the electric drift velocity $\mathbf{u}$. Consider, then, our usual large-scale magnetic field and plasma configuration, in which the characteristic scale l of the plasma and magnetic field is large compared to the cyclotron radius R of the individual ions and electrons. In that case, the motion of the individual charged particle is accurately described by the familiar guiding center approximation, recognizing that the primary transverse motion of a particle of mass M and charge q is the

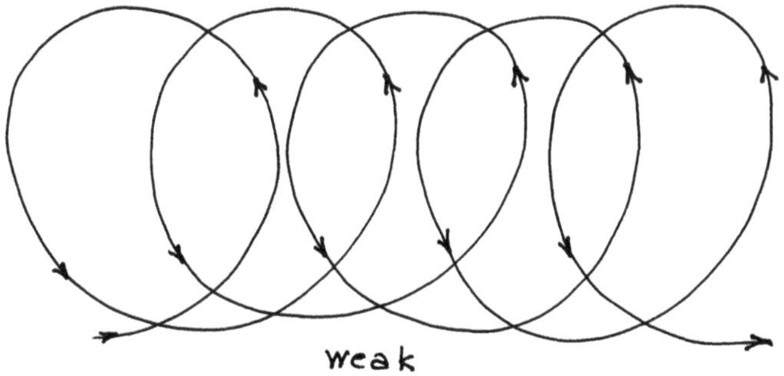

Fig. 9.3 A sketch of the path of a charged particle projected on the plane perpendicular to the local magnetic field. The magnetic field is slightly stronger on the upper side of the particle orbit and slightly weaker on the lower side, causing a variation in the cyclotron radius and a consequent drift of the guiding center of the orbit.

cyclotron motion in a circle of radius $R = w_n/\Omega$, with $\Omega = qB/Mc$, for a particle with thermal velocity w_n perpendicular to **B**. There is usually a small, but nonvanishing, gradient $O(B/l)$ in the field strength across the diameter $2R$ of the circular cyclotron trajectory. Thus, the curvature of the otherwise perfectly circular trajectory varies slightly from one side to the other, causing the circular trajectory to drift slowly, as may be seen from the sketch in Fig. 9.3. Define l as

$$\frac{1}{l} = \frac{|\nabla_n B|}{B} \tag{9.41}$$

where ∇_n represents the perpendicular component of the gradient of B. The mean drift velocity $\mathbf{v}_G$ of the circular trajectory is given by

$$\mathbf{v}_G = \frac{Mw_n^2 c}{2qB^4}\mathbf{B} \times \nabla\left(\frac{B^2}{2}\right) \tag{9.42}$$

The drift speed can then be written

$$\mathrm{v}_G = \frac{R}{2l} w_n \tag{9.43}$$

along the local contour of constant B.

Now the same particle, moving with velocity w_s along a field line that has curvature K, experiences a centrifugal force $F = Mv_s^2 K$, causing the particle to drift perpendicular to both **F** and **B** at the rate

$$\begin{aligned} \mathbf{v}_C &= \frac{F}{M\Omega} \\ &= \frac{w_s^2 K}{\Omega} \end{aligned} \tag{9.44}$$

such that the Lorenz force of the drift opposes and cancels the centrifugal force **F**. Formally,

$$\mathbf{v}_C = \frac{Mw_s^2 c}{qB^4}\mathbf{B} \times [(\mathbf{B} \cdot \nabla)\mathbf{B}] \tag{9.45}$$

where now w_s^2 represents the mean square thermal velocity parallel to the magnetic field.

Now the particle velocity perpendicular to **B** consists of the motion $\mathbf{u}_P + \mathbf{v}_G + \mathbf{v}_C + \mathbf{v}_P$ of the center of the cyclotron circle—the guiding center—plus the circular cyclotron motion around the center with the thermal speed w_n in the plane perpendicular to **B**. The diagonal component of the pressure tensor p_n as a consequence of the thermal motions perpendicular to **B** follows from eqn. (8.7) as

$$p_n = \frac{1}{V}\sum \frac{1}{2} M w_n^2 \tag{9.46}$$

where the sum is over all particles in a small volume V and M is the generic mass of the ion or electron. It is evident, then, that p_n represents the total thermal energy per unit volume in the two directions perpendicular to **B**. Similarly, the diagonal component of the pressure tensor p_s arising from the thermal motions w_s parallel to **B** follows as

$$p_s = \frac{1}{V}\sum M w_s^2 \tag{9.47}$$

It is evident that p_s represents twice the total kinetic energy of the thermal motions along **B**.

The next step is to compute the net current density j_n perpendicular to **B** by summing over all the particle motions, given by eqns. (9.40), (9.42), and (9.45), plus the current associated with the cyclotron motion of the individual ions and electrons in the presence of density and temperature gradients. This involves some messy geometrical constructions (Parker 1957a), which need not be repeated here. The essential feature of the sum over ions and electrons is that the thermal motions sum to p_n or to p_s in each contributing term, with the result that

$$\mathbf{j}_\perp = \frac{c}{B^2}\mathbf{B} \times \left\{ \nabla p_n + \left[\frac{4\pi(p_s - p_n)}{B^2}\right]\frac{(\mathbf{B}\cdot\nabla)\mathbf{B}}{4\pi} + N(M+m)\frac{d\mathbf{u}}{dt} \right\} \tag{9.48}$$

Substituting this current into Ampere's equation (3.11) leads to

$$N(M+m)\frac{d\mathbf{u}}{dt} = -\nabla_n\left(p_n + \frac{B^2}{8\pi}\right) + \frac{[(\mathbf{B}\cdot\nabla)\mathbf{B}]_n}{4\pi}\left(1 - \frac{p_s - p_n}{B^2/4\pi}\right) \tag{9.49}$$

where the subscript n denotes the component perpendicular to **B**. We see that this equation is the MHD momentum equation for the bulk motion perpendicular to **B** with the additional term $4\pi(p_s - p_n)/B^2$. That term represents the centrifugal force of any excess thermal motions along the curved field lines, pushing outward on the curved field in opposition to the tension in the field that pulls inward. The effect vanishes for isotropic thermal motions, $p_s = p_n$.

The derivation of this momentum equation directly from Newton's equation, rather than through application of Ampere's equation, can be found in Bittencourt (1986).

The essential point here is that the electric current automatically satisfies Ampere's law provided only that the individual thermal motions and the bulk motion satisfy Newton's equations of motion. So, as remarked earlier, Newton and Maxwell are automatically satisfied because they are both fundamental laws of nature.

We constructed this calculation many years ago (Parker 1957) in the hope of finding some circumstance under which Ampere's equation would not be satisfied automatically, thereby creating a nonnegligible displacement current, $\partial\mathbf{E}/\partial t$, and a runaway electric field. It would have been interesting, but the result was what the reader sees here, viz. complete compatibility within the confines of ordinary HD and MHD. It was that calculation that opened our eyes to the universal applicability of HD and MHD, even when thermal anisotropy is present. It was an essential first step in recognizing the

HD origin of the solar corpuscular radiation (Parker 1958, 1963a) in the expanding corona of the Sun. The concept of corpuscular radiation, of mysterious origin at the Sun, was replaced with the hydrodynamic concept of the coronal expansion providing the solar wind.

The bottom line is that the current automatically satisfies Ampere's law. Concern about whether the current accomplishes that task is a nonproblem in a highly conducting collision-dominated plasma, in a highly conducting partially ionized gas, or in a collisionless plasma.

9.7 Particle Motion Along B

The particle motion along **B** is free except for the mirror force, arising from the convergence or divergence of the magnetic field in the direction of **B**. With the finite cyclotron radius $R = Mcw_n/qB$ of a particle with mass M, charge q, and velocity w_n perpendicular to **B**, it is obvious that the field direction at the cyclotron orbit of the individual particle is different from the direction of the field at the guiding center of the orbit, as

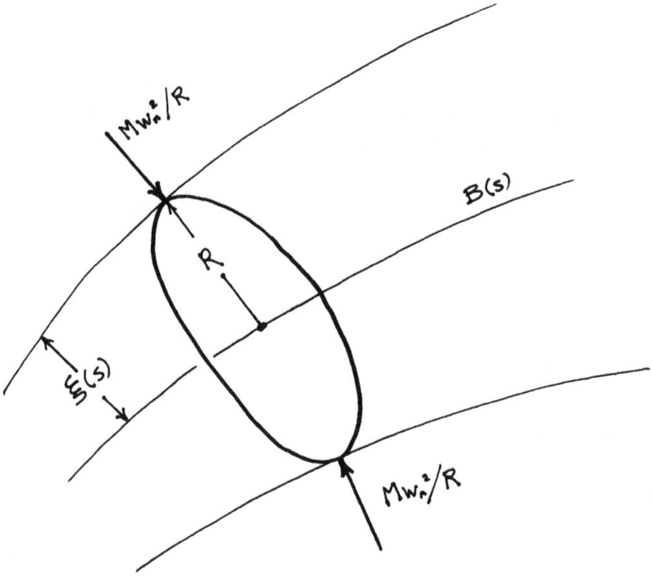

Fig. 9.4 A schematic drawing of the cyclotron orbit of a particle with charge q and mass M moving perpendicular to the magnetic field B with velocity w_n. The magnetic field strength $B(s)$ varies with distance s along the field, and the cyclotron radius is $R(s) = Mw_nc/qB(s)$.

is readily seen from Fig. 9.4. Conservation of magnetic flux ($\nabla \cdot \mathbf{B} = 0$) requires that the cross-sectional area $A(s)$ of any flux bundle multiplied by the field intensity $B(s)$ does not vary with distance s measured along the field. That is to say, the total magnetic flux, $\Phi = A(s)B(s)$, is constant along the bundle. Consider, then, the flux bundle with radius $\xi(s)$, so that $A(s) = \pi\xi(s)^2$. It follows that

$$\frac{d\xi}{ds} = -\frac{\xi}{2B}\frac{dB}{ds} \tag{9.50}$$

So if we consider the flux bundle enclosed by the cyclotron orbit of radius R, it follows that the inclination of the field at the orbit to the field at the center of the orbit is

$$\frac{d\xi}{ds} = -\frac{R}{2B}\frac{dB}{ds} \tag{9.51}$$

The inclination is small, of course, because $B(s)$ varies on the large scale $l \gg R$. The Lorentz force on the particle, opposing the centrifugal force $w_n{}^2/R$, is inclined to the radial direction by the small angle $d\xi/ds$, so it has the component

$$\begin{aligned} F_s &= -\frac{Mw_n^2}{R}\frac{d\xi}{ds} \\ &= -\frac{Mw_n^2}{2B}\frac{dB}{ds} \end{aligned} \tag{9.52}$$

in the direction of the field. This is called the *mirror force*, repelling the particle along the field from regions of strong B. It follows that the equation of motion for the particle velocity $\mathbf{u}_s$ along the field can be written

$$\frac{du_s}{dt} = -\frac{w_n^2}{2B}\frac{dB}{ds} \tag{9.53}$$

More generally,

$$\frac{d\mathbf{u}_s}{dt} = -\frac{w_n^2}{2B^4}\mathbf{B}\{\mathbf{B} \cdot [(\mathbf{B} \cdot \nabla)\mathbf{B}]\} \tag{9.54}$$

This result is sometimes misinterpreted to imply that the mirror force causes a reduction in particle density in regions of large $B(s)$. If particles are reflected from the region of dense field, surely the density is smaller there. In fact for an isotropic thermal distribution, the mirror force repels particles from a region of increased field strength in just the right amount to compensate for the increase of particle density that would arise from the funneling compression effect of the converging field lines. As we shall see a little farther along, there is a decrease in density of particles only if the thermal velocities have a preponderance of perpendicular motion w_n.

Note, then that the kinetic energy of the particle motion along the field with velocity u_s varies according to

$$\frac{d}{ds}\left(\frac{1}{2}Mu_s^2\right) = -\frac{1}{2}Mw_n^2\frac{1}{B}\frac{dB}{ds} \tag{9.55}$$

where we have written

$$\frac{du_s}{dt} = \frac{du_s}{ds}\frac{ds}{dt} = u_s\frac{du_s}{ds} = \frac{d}{ds}\left(\frac{1}{2}u_s^2\right)$$

in the presence of a magnetic field that is stationary in time. Now the total kinetic energy of the particle is constant in time, so that

$$w_n^2(s) + u_s^2(s) = w_n^2(0) + u_s^2(0)$$

Then, noting that $w_n^2(s)/B(s)$ is an invariant (see appendix E), it follows that

$$w_n^2(s) = w_n^2(0)\frac{B(s)}{B(0)}$$

With these conditions, we may rewrite eqn. (9.55) and integrate the result to give

$$u_s^2(s) = u_s^2(0) + w_n^2(0)\left[1 - \frac{B(s)}{B(0)}\right]$$

Note, then, that the velocity $u_s(s)$ declines to zero, and the particle is reflected back along the field, at the *mirror point* where $B(s)$ increases to

$$\frac{B(s)}{B(0)} = \frac{w_n^2(0) + u_s^2(0)}{w_n^2(0)} \tag{9.56}$$

Now, the collective bulk motion of the collisionless particles along the stationary field cannot be treated in the simple manner employed in constructing eqns. (8.5), (8.12), and (8.15) because now the particles are deflected between the parallel and perpendicular direction by mirroring in the variations *dB/ds* of the field intensity. So we turn to a more detailed approach, for which the collisionless Boltzmann equation is a convenient tool. The calculation is carried out in the local frame of reference moving with the mean bulk plasma velocity v_n perpendicular to the time-independent, large-scale field $B(s)$. Denote the total particle velocity by **u** and introduce the pitch angle θ between **u** and **B**. It follows that the velocity along the field is $u_s = u \cos\theta$ and the velocity perpendicular to the field is $w_n = u \sin\theta$. The Boltzmann equation describes the variation of the particle distribution function $F(s, \theta, t)d\theta$, giving the number of particles per unit volume in the interval $d\theta$.

The collisionless Boltzmann equation (cf. Parker 1957) can be written

$$\frac{\partial F}{\partial t} + u\cos\theta\frac{\partial F}{\partial s} + \frac{u\sin\theta}{2B}\frac{dB}{ds}\frac{\partial F}{\partial s} - \frac{u\cos\theta}{2B}\frac{dB}{ds}F = 0 \quad (9.57)$$

for particles with the constant velocity u in a stationary magnetic field $B(s)$. The number density of the particles is

$$N(s, t) = \int_0^{\pi} d\theta F(s, \theta, t) \quad (9.58)$$

The mean bulk velocity $v_s(s, t)$ along the field is given by

$$N(s, t)v_s(s, t) = \int_0^{\pi} d\theta F(s, \theta, t) u\cos\theta \quad (9.59)$$

The motion perpendicular to the field is the thermal cyclotron velocity w_n, given by

$$N(s, t)w_n(s, t) = \int_0^{\pi} d\theta F(s, \theta, t) u\sin\theta \quad (9.60)$$

Integrate the Boltzmann equation (9.57) over θ from 0 to π, obtaining

$$\frac{\partial N}{\partial t} + \frac{\partial}{\partial s}Nv_s + \frac{u}{2B}\frac{dB}{ds}\int_0^{\pi} d\theta\left(\frac{\partial F}{\partial \theta}\sin\theta - F\cos\theta\right) = 0$$

Integrating by parts leads to

$$\frac{\partial N}{\partial t} + B(s)\frac{\partial}{\partial s}\frac{Nv_s}{B(s)} = 0 \tag{9.61}$$

To see the geometrical basis for this relation, note that the cross-sectional area $A(s)$ of each elemental flux bundle varies inversely with $B(s)$, so that

$$\frac{\partial N}{\partial t} + \frac{1}{A(s)}\frac{\partial}{\partial s}A(s)Nv_s = 0 \tag{9.62}$$

and we have the expected statement that the time rate of change of the number of particles $A(s)N(s)$ per unit length of flux tube is equal to the negative divergence of the flux of those particles along the field. Thermal anisotropy appears only implicitly through variations in v_s, which we take up next.

To compute the momentum equation for the bulk flow of particles Nv_s along the field, multiply the Boltzmann equation (9.57) by $Mu\cos\theta$, and note that

$$\begin{aligned}\int_0^{\pi} d\theta F(s, \theta, t)Mu^2\sin^2\theta &= NMw_n^2 \\ &= p_n\end{aligned}$$

The kinetic energy density of the flow parallel to **B** is the energy density $\frac{1}{2}NMv_s^2$ of the bulk flow plus the kinetic energy of the mean square thermal velocity w_s parallel to the magnetic field, $\frac{1}{2}NMw_s^2$, with the diagonal component of the pressure tensor $p_s = NMw_s^2$. Thus,

$$\int_0^{\pi} F(s, \theta, t)Mu^2\cos^2\theta = NMv_s^2 + p_s$$

The integral of $Mu\cos\theta$ over the Boltzmann equation yields

$$\frac{\partial}{\partial t}NMv_s + \frac{\partial}{\partial s}NMv_s^2 - \frac{1}{B}\frac{dB}{ds}NMv_s^2 = -\frac{\partial p_s}{\partial s} + \frac{p_s - p_n}{B}\frac{dB}{ds} \tag{9.63}$$

upon integrating by parts again the term in $\partial F/\partial\theta$. Then multiply eqn. (9.61) by Mv_s and subtract from eqn. (9.63), with the result that

$$NM\left(\frac{\partial v_s}{\partial t} + v_s\frac{\partial v_s}{\partial s}\right) = -\frac{\partial p_s}{\partial s} + \frac{p_s - p_n}{B}\frac{dB}{ds} \tag{9.64}$$

for the component of the momentum equation parallel to **B**.

For thermal isotropy, $p_s = p_n$, eqn. (9.64) becomes the familiar Euler equation for fluid flow. Note, then, that if $p_n > p_s$, an increase in $B(s)$ along the field introduces a net decrease in the bulk flow velocity as a consequence of the repulsive mirror force dominating over the convergence of the field lines. On the other hand, if $p_s > p_n$, the converging field lines dominate over the mirror force and there is a net increase in the momentum density of the plasma flow into the region of increasing field $B(s)$. The mean bulk velocity also increases with increasing B, because the particles with the larger pitch angles and, hence, smaller v_s, are reflected back out of the region of larger B.

As a final comment, suppose that the distribution function $F(s, \theta, t)$ has the form $f(s) \sin^\alpha \theta$, with thermal isotropy for $\alpha = 1$. It follows from the Boltzmann equation that

$$\frac{df}{ds} + \frac{\alpha - 1}{2} \frac{f}{B} \frac{dB}{ds} = 0$$

from which it follows that

$$f(s) \propto B(s)^{(1-\alpha)/2} \tag{9.65}$$

Thus, for isotropy, the mirror force and the converging field are in balance and the particle density $f(s)$ is uniform along the field. On the other hand, when thermal velocities perpendicular to the field predominate, then $\alpha > 1$ and the density declines into a region of stronger field. Note again that this situation ($p_s > p_n$) is sometimes achieved in the solar wind, presumably as a consequence of selective wave heating of w_n in the plasma. In contrast, in the vicinity of the orbit of Earth, the principal expansion of the solar wind is in the two nonradial (transverse) directions relative to the approximately radial magnetic field. Thus, in the absence of significant wave heating, $p_s > p_n$ so that $\alpha < 1$, and the particle density would increase into a region of stronger field because the effect of the converging field lines dominates over the mirror force.

9.8 Time-varying Magnetic Field

The next step is to treat the variations of p_s and p_n in the presence of inhomogeneous bulk motion in a time-varying magnetic field. We need tractable equations for p_s and p_n in terms of N, $\mathbf{B}$, v_s, and v_n. Standard procedure is to attempt to solve the Boltzmann equation simultaneously with Maxwell's equations. However, the mathematical complexity (cf. Wu 1966) greatly limits progress in this direction. We think it is more informative in the spirit of these "conversations" to treat the collisionless case in the Chew-Goldberger-Low (1956) double adiabatic formalism for computing p_s and p_n in plasmas that are more or less isothermal.

Imagine an ion or electron moving along a field line of a large-scale magnetic field $B(s)$ that varies only slowly with the passage of time. There are several invariant quantities that come to mind, beginning with the transverse invariant

$$\frac{w_n(s)^2}{B(s)} = \text{constant}$$

demonstrated in appendix E. Then imagine that the particle moves freely along the field for distances comparable to the characteristic scale l of the magnetic field. The field varies strongly over the distance l, so it is reasonable to suppose that particle mirroring takes place, with the longitudinal invariant applying to particles reflected back and forth along the magnetic field across the length l of the region, giving

$$lw_s = \text{constant}$$

A flux bundle of cross-sectional area $A(s)$ satisfies the flux invariant

$$B(s)A(s) = \text{constant}$$

along the field. Conservation of particles suggests that there should be some such loose relation as

$$N(s)A(s)l = \text{constant}$$

Using these four relations, eliminate l and $A(s)$. Recalling that $p_n = \frac{1}{2} NMw_n^2$ and $p_s = NMw_s^2$, it is readily shown that p_n/NB and p_sB^2/N^3 are constants in each flux bundle, i.e.,

$$\frac{dp_n}{dt} = \frac{d}{dt}NB \qquad \frac{dp_s}{dt} = \frac{d}{dt}\left(\frac{N^3}{B^2}\right) \tag{9.66}$$

In as much as these two expressions for p_n and p_s depend on conditions along the length l, they do not really apply to an individual point. They are a statement of average conditions along the field. So we might say that they are advisory and informative. Bittencourt (1986, pp. 314–319) gives a more detailed derivation of the two expressions, and provides some elementary applications to one- and two-dimensional compression. The results are our eqn. (8.18) for one dimensional compression, and eqn. (8.22) with $\tau_1 = \tau_2$, $\tau_3 = \infty$ for isotropic two-dimensional compression.

As already noted, the formally correct approach is to plunge into the plasma kinetic treatment of the problem, which is intractable in the general case and can be properly used only in simple cases. Landau damping of compressional waves appears in the proper treatment, of course.

Now the Chew, Goldberger, Low double-adiabatic concept can be taken one step farther, to recognize the consequences of interparticle collisions and wave scattering. A linear scattering term can be introduced to represent the collisonal trend toward thermal isotropy, yielding

$$\frac{dp_n}{dt} = \frac{d}{dt} NB + \frac{p_s - p_n}{\tau} \tag{9.67}$$

$$\frac{dp_s}{dt} = \frac{d}{dt}\left(\frac{N^3}{B^2}\right) - \frac{p_s - p_n}{\tau} \tag{9.68}$$

where τ is the characteristic scattering time. Wave scattering may be similarly introduced, taking a variety of forms depending upon the nature of the waves. For instance, a simple trend toward thermal isotropy is possible, as well as preferential pumping of p_n, as in the solar wind.

Overall, it must be appreciated that particles are scattered by Coulomb collisions with other particles, by waves generated locally by thermal anisotropies, and by waves propagating into the region from their place of origin elsewhere. Hence, the degree of anisotropy of the thermal motions often cannot be calculated with anything approaching certainty. So, in practice, we examine simple models to investigate the theoretical possibilities, beginning with the basic cases of thermal isotropy and a scalar pressure, and with the expectation that in most cases the actual degree of anisotropy does not dominate the bulk dynamics in any important way.

9.9 Comments

Let us take stock as to where this discussion leaves us in our investigation of the dynamics of the plasmas and magnetic fields in the cosmos. A number of remarks come to mind. First of all, let us be clear on when a plasma is collisionless or collision-dominated. The concept of bulk flow **v** over a scale l applies when l is large compared to the thermal ion cyclotron radius and the ion inertial length. The scale l provides a characteristic dynamical time l/v, and that characteristic time may be long or short compared to the characteristic time in which interparticle Coulomb collisions and wave-particle scattering drive the plasma toward thermal isotropy. If l/v is shorter, then the plasma is effectively collisionless. The dynamics follows along the lines describe in the foregoing sections. If longer, then the plasma may be considered collision-dominated, and its dynamical behavior approximated with a simple scalar pressure p, described by the familiar energy equation (8.46). The next level of investigation, then, uses the isotropic model to estimate the degree of thermal anisotropy and the net heat input from wave-particle scattering, which can be accomplished only to the degree that the plasma instabilities and the wave spectrum are understood. The solar wind out here in space at a distance of 1 AU is an example of a plasma that is neither completely collision dominated nor collision free. The high ion temperature, with p_n sometimes as large as p_s, and both much above the expected adiabatic values, shows the importance of a distributed heat input. The Chew-Goldberger-Low double-adiabatic approach is useful here as a general diagnostic, giving some idea of the variation of the thermal anisotropy from the bulk motion alone. Then it must be appreciated that the heat input involves waves generated closer to the Sun, possibly in the corona, so the problem becomes nonlocal. The interaction of the solar wind with the magnetospheres of the planets drives an overall convection of the magnetospheres, which is another nonlocal problem, depending strongly on the state of the plasma and magnetic field of the wind sweeping by the magnetosphere and the effective dynamical friction and drag that the wind is able to exert on the magnetosphere. The convection is opposed, then, by the friction and drag that the nonconducting atmosphere exerts on the base of the convecting magnetosphere.

It is curious, then, that the large-scale magnetic fields in the cosmos, described by the linear induction equation (9.2), provide such complex dynamics through coupling to the electrically conducting plasmas. The plasma contributes much of the complexity, but the quadratic form of the Maxwell stress tensor of the magnetic field introduces a whole new world of HD. Such phenomena as rapid reconnection, the aurora, the fibril structure of the magnetic fields at the surface of the Sun, the X-ray corona of the Sun, the expanding quiet corona and the solar wind, the magnetic substorm here at Earth, and the particle acceleration in rapid reconnection and in shock fronts are all consequences of the dynamics produced by magnetic fields in a plasma. So it seems appropriate to explore one of the more exotic aspects of the Maxwell stress tensor, responsible for rapid reconnection of magnetic field at incipient singularities forming in the untidy magnetic field line topology arising in the chaotic universe.

10 Singular Properties of the Maxwell Stress Tensor

10.1 Magnetic Equilibrium

We come now to the magnetic fields embedded in a highly conducting plasma and subject to mixing and interlacing of the field lines by the swirling plasma. For the fact is that the Maxwell stress tensor,

$$M_{ij} = -\delta_{ij}\frac{B^2}{8\pi} + \frac{B_iB_j}{4\pi} \tag{10.1}$$

for a magnetic field with an "untidy" field line topology has the curious property of creating internal surfaces of tangential discontinuity, i.e., current sheets, as the magnetic field relaxes to equilibrium (Parker 1972). Thus, while the dynamical swirling of the magnetic field defines a large-scale mixing length l, the magnetic stresses acting alone cause the magnetic field gradients to steepen without limit when the field is allowed to relax to the lowest available energy state.

To fix ideas, consider the uniform magnetic field B extending in the z direction from $z = 0$ to $z = L$ through an ideal infinitely conducting, incompressible fluid. There is an infinitely conducting rigid endplate at $z = 0$ into which the magnetic field is permanently fixed. The endplate at $z = L$ is similarly infinitely conducting, but it is subject to arbitrary incompressible two-dimensional mechanical mapping in its own plane. At time $t = 0$ we switch on the arbitrary two-dimensional fluid motion

$$v_x = +kz\frac{\partial\psi}{\partial y} \qquad v_y = -kz\frac{\partial\psi}{\partial x} \qquad v_z = 0 \tag{10.2}$$

including the endplate at $z = L$, where $\psi = \psi(x, y, kzt)$. We choose $\psi(x, y, kzt)$ to be an arbitrary bounded, continuous, smooth, n-times differentiable function of each of its arguments. That is to say, ψ is a well-behaved, arbitrarily complicated function of its arguments. We can see from eqn. (10.2) that v_x and v_y represent an arbitrary swirling of the plasma with the motion fed in at $z = L$ and migrating downward into $0 < z < L$ with the passage of time. Figure 10.1 is a sketch of the interlacing of the field lines after a time t, when the field is (Parker 1986)

$$B_x = +Bkt\frac{\partial\psi}{\partial y} \qquad B_y = -Bkt\frac{\partial\psi}{\partial x} \qquad B_z = B \tag{10.3}$$

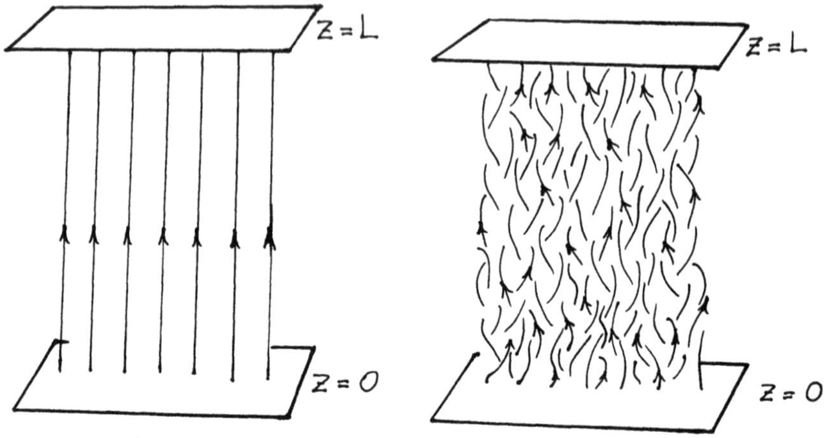

Fig. 10.1 A schematic drawing of the interlaced magnetic field described by eqn. (10.3), created from the initially the uniform magnetic field B by the incompressible swirling of the plasma described by eqn. (10.2)

After a time $t = \tau$, the fluid motion is switched off and the end plate at $z = L$ is held fixed while the fluid throughout $0 < z < L$ is released so that the field can relax to the lowest available energy state for the existing frozen in magnetic field topology. The fluid pressure p is maintained constant and uniform at both end plates, so that the pressure p is uniform throughout $0 < z < L$. We add a small viscosity to dissipate the motions of the fluid so that the relaxation process has a finite characteristic time.

Now both ends of the field are connected into the endplates at $z = 0$ and $z = L$, and the field is embedded in an infinitely conducting fluid throughout $0 < z < L$ so the field lines do not break or reconnect. There is clearly a lowest available energy state, i.e., a stable equilibrium. Equally clearly, the interlacing of the field lines introduced by $\psi(x, y, kzt)$ is arbitrary and subject to unlimited complication, i.e., diverse arbitrary patterns of interlacing and untidiness along the length L of the field. The length L may be as large as we please with any number of successive different patterns of interlacing along the field, because the interlacing is a simple matter of prescribing fluid motion.

The curious property of M_{ij}. appears when we find that the mathematical properties of the equilibrium condition, $\partial M_{ij}/\partial x_j = 0$, are unable to accommodate the arbitrary interlacing of the field lines. This is a seeming contradiction. Yet there are no contradictions in nature. Contradictions arise only when the theoretician has made an unjustified assumption. So the next step is to determine where the analysis has gone wrong. The answer turns out to be, of course, that we have supposed that the final equilibrium field configuration is continuous everywhere in the region. That overly constrains the mathematics to the point that there are no analytic solutions for almost all field line interlacing. Introducing discontinuities, in the form of surfaces of tangential discontinuity, i.e., current sheets, provides an extra degree of freedom for the mathematics and the arbitrary interlacing can then be accommodated. It follows that the equilibrium state of almost all field line topologies involves surfaces of tangential discontinuity.

To take the development one step at a time, the uniform fluid pressure means that the magnetic stresses are in equilibrium with themselves,

$$\frac{\partial M_{ij}}{\partial x_j} = 0$$

so that the tension $B^2/4\pi$ along the curving field lines balances the gradient of the isotropic pressure $B^2/8\pi$. This condition can also be written

$$(\nabla \times \mathbf{B}) \times \mathbf{B} = 0$$

so that the tension–pressure balance becomes

$$\nabla \frac{B^2}{8\pi} = \frac{(\mathbf{B}\cdot\nabla)\mathbf{B}}{4\pi}$$

The general condition on $\mathbf{B}$ follows as

$$\nabla \times \mathbf{B} = \alpha(\mathbf{r})\mathbf{B} \tag{10.4}$$

where the "torsion" coefficient $\alpha(\mathbf{r})$ is an unspecified scalar function of position.

First of all, it must be appreciated that eqn. (10.4) is not a linear equation for $\mathbf{B}$, because $\alpha(\mathbf{r})$ depends on $\mathbf{B}(\mathbf{r})$. Then note that the curl of eqn. (10.4) can be written

$$\nabla^2 \mathbf{B} + \alpha^2 \mathbf{B} = \mathbf{B} \times \nabla\alpha \tag{10.5}$$

The term $\nabla^2 \mathbf{B}$ indicates a quasilinear elliptic equation. That is to say, eqn. (10.4) has two families of complex characteristics, and specification of $\mathbf{B}$ on any enclosing surface uniquely determines $\mathbf{B}$ throughout the enclosed volume. However, the situation is more complicated than that, as may be seen from the fact that the divergence of eqn. (10.4) becomes

$$\mathbf{B}\cdot\nabla\alpha = 0 \tag{10.6}$$

This equation asserts that $\alpha(\mathbf{r})$ is constant along each field line. Obviously, the field lines represent a family of real characteristics of the equilibrium equation (10.4). Specification of $\alpha(\mathbf{r})$ on a given field line in no way constrains $\alpha(\mathbf{r})$ on any adjacent field line. That is to say, there is no requirement that $\alpha(\mathbf{r})$ is continuous from one field line to the next. The essential point is, then, that the equilibrium equation is not purely elliptic. It has mixed characteristics, with two families of complex characteristics and one family of real characteristics, quite unlike the familiar purely elliptic equations and purely hyperbolic equations that represent most of classical physics. So we must understand the physical implications of the unfamiliar mathematical properties of eqn. (10.4) (see discussion in Parker 1979, chapter 14; Parker 1994).

The constraint indicated by eqn. (10.6) indicates an immediate problem, viz. the torsion coefficient $\alpha(\mathbf{r})$ is constant along each individual field line of $\mathbf{B}$ all the way from $z = 0$ to $z = L$ no matter what the nature of the successive patterns of swirling and interlacing through which the individual field line extends. A right-hand spiral at one location along the field line must be accommodated by the same torsion $\alpha(\mathbf{r})$ as the left-hand spiral in a wholly different topological swirl somewhere else along the field line. It is not obvious that this is possible, so we begin with a brief study of the physical significance of $\alpha(\mathbf{r})$.

It is evident from eqn. (10.4) that $\alpha(\mathbf{r})$ is a measure of the strength of $\nabla \times \mathbf{B}$ relative to $\mathbf{B}$. This can be expressed in terms of the local magnetic circulation around a small closed contour C circling any given slender flux bundle. The circulation is defined as

$$\Gamma \equiv \oint_C d\mathbf{s} \cdot \mathbf{B}$$

Stokes's theorem converts this to the integral over the area S enclosed by C,

$$\Gamma = \int_S d\mathbf{S} \cdot \nabla \times \mathbf{B}$$

$$= \int_S d\mathbf{S} \cdot \mathbf{B}\alpha(\mathbf{r})$$

If $\alpha(\mathbf{r})$ is a continuous function of position $\mathbf{r}$, then for a sufficiently small contour C, $\alpha(\mathbf{r})$ can be approximated by its value on the field line. The result is

$$\Gamma = \alpha\Phi \qquad (10.7)$$

where Φ is the magnetic flux through C,

$$\Phi = \int_C d\mathbf{S} \cdot \mathbf{B} \tag{10.8}$$

Equation (10.7) states that $\alpha(\mathbf{r})$ represents the magnetic circulation per unit flux. Thus, eqn. (10.4) asserts that the circulation per unit flux has the same value everywhere along each field line.

Note, then, that $\alpha(\mathbf{r})$ is also a measure of the magnetic torque transmitted along the field. In the simple case that the contour C is a circle of small radius a, the torque Θ can be written

$$\Theta = \frac{B}{4\pi} \int_{-a}^{+a} d\xi \int_{-(a^2-\xi^2)^{1/2}}^{+(a^2-\xi^2)^{1/2}} d\eta(\xi B_\eta - \eta B_\xi)$$

where ξ and η are local Cartesian coordinates in the plane perpendicular to $\mathbf{B}$ with the origin at the position of the given field line so that B_ξ and B_η vanish at the origin. Expand B_ξ and B_η about the origin and neglect terms third order and higher in ξ and η. The result is

$$\Theta = \frac{\alpha\Phi^2}{16\pi^2} = \frac{\alpha\Phi^2}{157.910}$$

where $\Phi = \pi a^2 B$. This relation states that the torque Θ transmitted by a given flux Φ is the same everywhere along the field line. The numerical proportionality coefficient is specific to the precise form of the area over which the torque is computed, of course, because the torque is a quadratic effect. Had we chosen a square of side $2a$ centered on the origin, the result would have been

$$\Theta = \frac{\alpha\Phi^2}{48\pi} = \frac{\alpha\Phi^2}{150.796}$$

with $\Phi = 4a^2$. The torque is determined by the radius of gyration of the area enclosed by the contour C. Indeed, had we chosen a rectangle with unequal sides, the torque would not be expressible in terms of α and Φ alone.

Finally, note that, while the ratio Θ/Φ^2 is a direct measure of α, it does not follow that the torque transmitted by a specific flux bundle extending along the given line of force is constant along the flux bundle. This follows from the fact that the shape of the cross section of the flux bundle varies along the length of the flux bundle for almost every flux bundle, given that $\nabla \times \mathbf{B}$ is generally nonvanishing. So it must be understood that α is a measure of the torque transmitted across a specified symmetric cross section everywhere along the field. And that does not constitute a flux bundle.

10.2 Calculation of the Equilibrium Field

Consider how we might compute the final unique equilibrium field produced by a specific choice of $\psi(x, y, kzt)$. For instance, does specification of the field on an end plate provide a complete and unique equilibrium solution throughout $0 < z < L$? The answer is affirmative, as may be seen from the fact that specification of $B_i(x, y, 0)$ on $z = 0$ allows the expression of B_i throughout $z > 0$ in ascending powers of z,

$$B_i(x, y, z) = B_i(x, y, 0) + \left(\frac{\partial B_i}{\partial z}\right)_0 z + \left(\frac{\partial^2 B_i}{\partial z^2}\right)_0 \frac{z^2}{2!} + \cdots$$

This is readily seen from the fact that the x component of the equilibrium equation (10.4) can be solved for $\partial B_y/\partial z$ at $z = 0$ given $B_i(x, y, 0)$. The y component can be solved in like manner for $\partial B_x/\partial z$ at $z = 0$. The z component yields α. Then solve $\nabla \cdot \mathbf{B} = 0$ for $\partial B_z/\partial z$ at $z = 0$. That takes the expansion through the terms first order in z. Then differentiate the expressions for $\partial B_i/\partial z$ with respect to z to obtain the second derivatives at $z = 0$, using the same expressions for $\partial B_i/\partial z$ to express everything in terms of $B_i(x, y, 0)$ again. This operation can be repeated as many times as desired, extending the expansion in z to all orders. It follows, then, that, barring singularities in the field, the complete equilibrium solution can, in principle, be extended throughout the entire region, $0 < z < L$. Similarly, an expansion in ascending powers of $L - z$ can be constructed in terms of $B_i(x, y, L)$ and extended downward all the way to $z = 0$.

This is no help in our quest for the equilibrium field, of course, because we cannot know the equilibrium field at either $z = 0$ or $z = L$ until the equilibrium problem is solved. We know only the field, given by eqn. (10.3), before the relaxation to equilibrium begins. We know from simple physical considerations that there is a final equilibrium field that follows uniquely from the field of eqn. (10.3). So the information specifying the final equilibrium field is preserved in some way through the relaxation process from eqn. (10.3) to eqn. (10.4). The one thing that is preserved during the relaxation is the topology of the field line interlacing. So, evidently, the essential information is contained in the field line topology. The field lines represent the family of real characteristics, of course, so it is the topology of the real characteristics that, in some way, determines the final equilibrium field. And that is the basic nature of the mathematics of the equilibrium, field. The topology of the real characteristics of the final equilibrium solution of eqn. (10.4) is identical with the topology of the magnetic field of eqn. (10.3). This fact, that the invariant topology of the interlacing is the essential determining feature of the final equilibrium, plays a central role in the discussion in the next section.

10.3 Equilibrium in Stretched Field

Let us try another tack. The essential features of the topology of the field lines are preserved if we stretch the region of field by a large factor $Q \gg 1$ in the z direction. So we introduce the dilatation mapping that each point z goes over into the new coordinate $Z(z)$, where $Z(z) = Qz$. The z component of the field is not greatly affected by this stretching operation, but obviously the two transverse components, B_x and B_y, are expanded by the factor Q and diminished in intensity by the factor $1/Q$. This does not linearize the equilibrium equation (10.4), but it simplifies the form, reducing eqn. (10.4) to the more familiar equation of hydrodynamic vorticity in two-dimensional flow of an ideal inviscid incompressible fluid. We will take advantage of the fact that the vorticity equation has been studied in depth.

Suppose that l represents the characteristic scale of variation of the initial field of eqn. (10.3) in each of the three directions, x, y, and z. Then after the dilatation the characteristic scale in the z direction is Ql, with the transverse scale (x and y) still given by l. Thus, following the dilatation the gradient in the z-direction is written $\partial/\partial Z$, which is now small, $O(\varepsilon)\partial/\partial x$ and $O(\varepsilon)\partial/\partial y$, where for convenience we let $\varepsilon = 1/Q \ll 1$. Write $\partial/\partial Z = \varepsilon\partial/\partial\varsigma$ so that in terms of the new coordinate $\varsigma = \varepsilon Z$, the derivative $\partial/\partial\varsigma$ is of the same order as $\partial/\partial x$ and $\partial/\partial y$. Write the magnetic field in the form

$$B_x = \varepsilon B b_x \qquad B_y = \varepsilon B b_y \qquad B_z = B(1 + \varepsilon b_z)$$

where B is the initial uniform field strength and remains as a useful measure of the field. The torsion coefficient α is stretched along the field in the z direction and reduced to the same degree as the transverse field components, so that we write $\alpha = \varepsilon a$. The components of the equilibrium eqn. (10.4) become

$$\frac{\partial b_z}{\partial y} - \varepsilon\frac{\partial b_y}{\partial \varsigma} = +\varepsilon a b_x \tag{10.10}$$

$$\varepsilon\frac{\partial b_x}{\partial \varsigma} - \frac{\partial b_z}{\partial x} = +\varepsilon a b_y \tag{10.11}$$

$$\frac{\partial b_y}{\partial x} - \frac{\partial b_x}{\partial y} = a(1 + \varepsilon b_z) \tag{10.12}$$

It is evident from eqns. (10.10) and (10.11) that b_z is small $O(\varepsilon)$ compared to b_x and b_y. Thus, the condition $\nabla \cdot \mathbf{B} = 0$ reduces to

$$\frac{\partial b_x}{\partial x} + \frac{\partial b_y}{\partial y} = O(\varepsilon^2) \tag{10.13}$$

Neglecting terms second order in ε compared to one, it follows that b_x and b_y can be expressed in terms of a flux function $\theta\ (x, y, \varsigma)$, with

$$b_x = +\frac{\partial \theta}{\partial y} \qquad b_y = -\frac{\partial \theta}{\partial x} \tag{10.14}$$

Equation (10.12) becomes

$$a = -\left(\frac{\partial^2 \theta}{\partial x^2} + \frac{\partial^2 \theta}{\partial y^2}\right) \tag{10.15}$$

Differentiate eqn. (10.10) with respect to y and eqn. (10.11) with respect to x. Subtract the second from the first, obtaining

$$\frac{\partial^2 b_z}{\partial x^2} + \frac{\partial^2 b_z}{\partial y^2} = +\varepsilon\left[\frac{\partial}{\partial x}\left(a\frac{\partial \theta}{\partial x}\right) + \frac{\partial}{\partial y}\left(a\frac{\partial \theta}{\partial y}\right)\right] \tag{10.16}$$

upon using eqn. (10.15) to eliminate θ in terms of a. Finally, eqn. (10.6) becomes

$$\frac{\partial a}{\partial \varsigma} = \frac{\partial \theta}{\partial x}\frac{\partial a}{\partial y} - \frac{\partial \theta}{\partial y}\frac{\partial a}{\partial x} \tag{10.17}$$

Equations (10.15) and (10.17) together provide a single nonlinear partial differential equation for θ.

More important, however, is the fact that eqns. (10.14), (10.15), and (10.17) are identical in form to the two-dimensional vorticity equation for an ideal incompressible inviscid fluid. For such a fluid the incompressibility, $\nabla \cdot \mathbf{v} = 0$, means that the velocity components can be expressed in terms a stream function $s(x, y, t)$,

$$v_x = +\frac{\partial s}{\partial y} \qquad v_y = -\frac{\partial s}{\partial x} \tag{10.18}$$

which is the precise counterpart of eqn. (10.14). Thus, the vorticity is

$$\omega = \frac{\partial v_y}{\partial x} - \frac{\partial v_x}{\partial y} \tag{10.19}$$

$$= -\left(\frac{\partial^2 s}{\partial x^2} + \frac{\partial^2 s}{\partial y^2}\right) \tag{10.20}$$

as the analog of eqn. (10.15). The vorticity equation, $d\omega/dt = 0$, becomes

$$\frac{\partial \omega}{\partial t} = \frac{\partial s}{\partial x}\frac{\partial \omega}{\partial y} - \frac{\partial s}{\partial y}\frac{\partial \omega}{\partial x} \tag{10.21}$$

in precise analogy to eqn. (10.17), with ω in place of a and t in place of ς. So the torsion a evolves with distance ς, i.e., with z, in the same way that the vorticity ω evolves with the passage of time t. The vorticity is constant along the worldline of each moving element of fluid, just as the torsion coefficient is constant along each field line.

This analogy is important because the vorticity equation has been studied in depth over the last century. The reader is referred to the review by Kraichan and Montgomery (1980). Thus, for instance, specification of $\omega(x, y, t)$ at any time $t = t_1$ uniquely determines $\omega(x, y, t)$ for all other t, just as specification of $\alpha(x, y, z_1)$ at some surface $z = z_1$ determines $\alpha(x, y, z)$ over all z. The studies of the vorticity equation show that the enstrophy ω^2 cascades with the passage of time to smaller and smaller scales, while the kinetic energy v^2 cascades to larger scales. The nonlinear character of the vorticity equation causes the cascades to be irreversible. That is to say, if the velocity is reversed, $s \rightarrow -s$, $\omega \rightarrow -\omega$, the right-hand side of eqn. (10.21) is unaffected, while the left-hand side changes sign, which is not the vorticity equation then. Only if time is also reversed does it remain the vorticity equation. It follows that the analogous magnetic equilibrium equation (10.17) has the same properties, the solutions showing an α^2 migrating irreversibly to smaller and smaller scales with increasing ς, while the energy of the transverse field, $b_x^2 + b_y^2$, migrates irreversibly to larger scales.

Now, we concluded in section 10.2 that the topology of the interlacing of the field is the essential determining feature of the equilibrium. The field line topology is preserved during the relaxation to equilibrium and ultimately determines the final unique equilibrium field configuration. Note, then, that the vorticity equation (10.21) decrees that the topology of the swirling pattern of flow at any give time t_1 determines the succeeding swirls all the way to $t = \infty$. Similarly, the swirling topological pattern of the field line interlacing at any given level ς_1 determines the succeeding interlacing patterns all the way to $\varsigma = L$.

The contradiction between the physics and the mathematical solutions is that the interlacing of the field, introduced initially by $\psi(x, y, kzt)$, generally represents a succession of independent and unrelated swirling patterns along the field. There is no systematic evolution of the interlacing patterns with increasing ς because we do not choose a $\psi(x, y, kzt)$ with those special evolutionary properties. We have in mind that in nature the statistical properties of the field line interlacing are the same throughout the entire interval $0 < z < L$.

In summary, then, the field line interlacing introduced by $\psi(x, y, kzt)$ does not have the special evolutionary form of the solutions of eqn. (10.17). On the other hand, the field lines are all tied at $z = 0$ and $z = L$, so the simple physics of the permanent magnetic field line connection guarantees the existence of a stable equilibrium, i.e., a lowest available energy state of the field, for any and all the successive unrelated patterns of field line interlacing that might be introduced by $\psi(x, y, kzt)$. On the other hand, the equilibrium equation (10.17) has no solutions that accommodate these arbitrary successions of interlacing patterns throughout $0 < z < L$. So where do we go from here? As already remarked, there are no contradictions in nature. So what is the error in our theoretical approach to the problem?

10.4 Resolving the Contradiction

The error in the development of the equilibrium field has been the conventional but unnecessary assumption that the equilibrium field is everywhere continuous. We have just seen how that leads to a qualitative contradiction for the interlaced magnetic field topologies. The innocent assumption of continuity ties us to the universal restriction of eqn. (10.6), that the local circulation of field around the mean field direction is the same along each individual field line no matter how the field line circulates around its neighboring field lines from one swirling pattern to the next. We pointed out in section 10.1 (Parker 1972) that this must lead to contradictions because a given field line may circulate first one way in one swirl and then the other way in another of the many unrelated swirls along the length of the field line. Suppose, then, that we admit the possibility of surfaces of tangential discontinuity (TD), i.e., current sheets, in the equilibrium magnetic field, sketched in Fig. 10.1. The field magnitude and pressure are continuous across the TD, but the field direction changes abruptly by some angle ϑ. A TD is allowed by the family of real characteristics, i.e., the field lines, on which specification of the field direction on a given field line does not constrain the field direction on neighboring lines.

The question is whether a TD can free us from the constraint of eqn. (10.6) so that a solution of the equilibrium equation (10.4) can accommodate the field line topology introduced by the arbitrary swirling of $\psi(x, y, kzt)$. The essential point is that the TD represents the surface of contact between two regions of continuous field on either side. Neither the field $\mathbf{B}$ nor $\nabla \times \mathbf{B}$ is defined on the TD. Hence, eqns. (10.4) and (10.6) are not applicable on the TD. The shear or torsion across the TD, described by the angle ϑ, is unrestricted by eqn. (10.6). Hence, any shearing of the field required by the interlacing of the field lines, but not compatible with eqn. (10.6), is accommodated by the formation of a TD. It must be emphasized, then, the TD gets around the application of eqn. (10.6) because the TD is a true mathematical singularity, so that **B** is undefined on the TD (Parker 1972, 1994). That is to say, the associated current sheet is infinitely intense and has vanishing thickness.

Given that the final complete equilibrium exists, it follows that the relaxation to equilibrium is an asymptotic process. The longest available characteristic relaxation time is L/C, where C is the Alfvèn speed in the mean field B. In the relaxation process the Maxwell stresses in the magnetic field push the fluid in such a way as to steepen the field gradient in the direction perpendicular to the incipient TD. This represents the rapid reconnection phenomenon, in which fluid gradient becomes so steep that resistive, or other, dissipation, no matter how weak, can "eat" away and reconnect the field as rapidly as the Maxwell stresses steepen the gradient. So in the real world, the magnetic field finds itself in a continuing dissipative near-equilibrium. The rate of dissipation of magnetic energy into heat is not limited by the enormous passive resistive dissipation time l^2/η for the field as a whole, with the characteristic large scale l. Rather it is determined by the dynamics of the fluid driven by the Maxwell stresses of the nonequilibrium magnetic field toward forming a TD, on which we comment in section 10.6. It follows that a magnetic field subject to continuing convective interlacing of its field lines must eventually reach a limiting state in which the rate of magnetic energy dissipation at incipient TDs is equal to the rate at which the convective swirling of the field does work on the field.

Now it should not go unnoticed that there was initial skepticism concerning the general theorem that an untidy field topology leads to surfaces of tangential discontinuity (cf. Van Ballegooijen 1985; and the discussion in Low 1990). However, this basic singular condition seems to be mathematically inescapable, and there is growing observational evidence that the theorem is at work in the X-ray corona of the Sun, summarized at the end of section 10.6. We presume, therefore, that the theorem is active in the magnetic fields of other stars.

10.5 Formation of TDs

How is a TD formed? How do the field lines extend through a region of untidy topology? To explore these questions we turn to the optical analogy (Parker 1989a, b, 1991, 1994) described in chapter 5 by eqns. (4.3)–(4.11). The optical analogy states that a field line in a potential field ($\nabla \times \mathbf{B} = 0$) extends along the same path as an optical ray in an index of refraction proportional to the magnitude $|\mathbf{B}|$ of the field. Consider, then, any flux surface in a field described by the equilibrium eqn. (10.4). Draw any continuous curve H across the field. The field lines intersecting H form a surface, which is commonly called a *flux surface*. The flux surface represents a two-dimensional space. The space is generally not flat, i.e., it is usually non-Euclidean, but the essential point is simply that a field in three dimensions satisfying eqn. (10.4) is a potential field in any 2D flux surface. For, if the 2D field had a nonvanishing curl, that curl would be perpendicular to the flux surface. Equation (10.4) states that the $\nabla \times \mathbf{B}$ lies parallel to $\mathbf{B}$ on the flux surface. There is no perpendicular component of $\nabla \times \mathbf{B}$. So the field in the flux surface has no curl and can be represented by a scalar potential. The optical analogy applies, stating that a field line connecting point 1 to point 2 lies along the path in the flux surface that minimizes the integral of $|\mathbf{B}|$ from 1 to 2.

Consider the effect of a local maximum in the field magnitude somewhere along the path connecting points 1 and 2. Figure 10.2 is a sketch of the layout, showing a local enhanced field magnitude ΔB above the surrounding B with dimensions w and l. The local maximum lies a distance λ_1 from point 1 and a distance λ_2 from point 2, so that the total distance is $\lambda_1 + \lambda_2$ from point 1 to point 2. Assume that $w, l \ll \lambda_1, \lambda_2$. The field line, or ray path, has the option of passing along the shortest

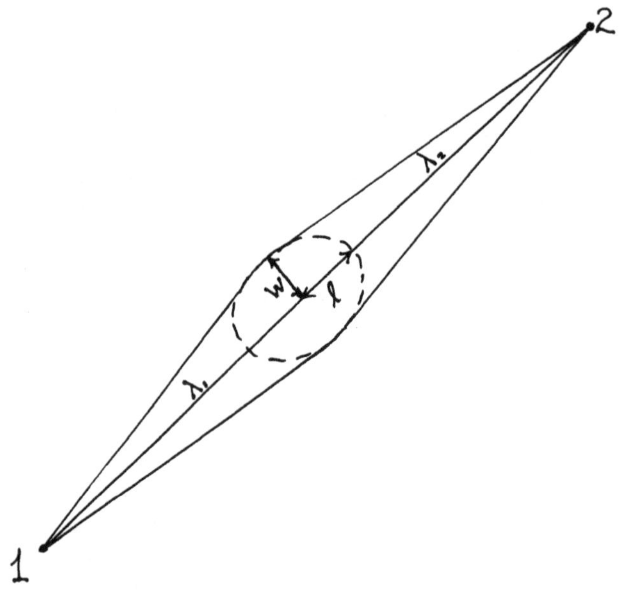

Fig. 10.2 A schematic drawing of the possible minimum paths of a field line connecting points 1 and 2 when there is an intervening localized maximum in the field magnitude located between the end points 1 and 2.

path over the top of the maximum or going around on one side to avoid the increased field magnitude. Passing straight over the top adds $2l\,\Delta B$ to the integral of $|\mathbf{B}|$, while passing around to the side avoids ΔB, but is longer by $w^2(1/\lambda_1 + 1/\lambda_2)/2$. When $l\,\Delta B$ is sufficiently large, the path around the side minimizes the integral of $|\mathbf{B}|$. This occurs for

$$\frac{\Delta B}{|\mathbf{B}|} > \frac{w^2}{2l}\left(\frac{1}{\lambda_1} + \frac{1}{\lambda_2}\right) \tag{10.22}$$

which is small compared to one for $w, l \ll \lambda_1, \lambda_2$. In the simple case that $\lambda_1 = \lambda_2$ and $w = l$, this criterion reduces to

$$\frac{\Delta B}{|\mathbf{B}|} > \frac{w}{\lambda_1} \ll 1 \tag{10.23}$$

A small enhancement of field magnitude suffices to expel the field line from the local maximum ΔB. So the field lines avoid ΔB, creating a gap in the flux surface. In fact, the local ΔB causes a gap in a slab of field of finite thickness, shown schematically in Fig 10.3. The essential point is that the magnetic fields on either side of the slab press together through the gap. The two fields are generally not parallel, so they create a TD at the surface where they meet through the gap.

Figure 10.4 is a schematic drawing of two flux bundles pulling around each other, their tension creating a modest compression of the field where they meet. A TD would be created by the ΔB in the locally increased magnetic pressure where they come in contact.

The TDs form a structure of their own. For instance, they end only at the boundaries $z = 0, L$. However, one TD may intersect another. It

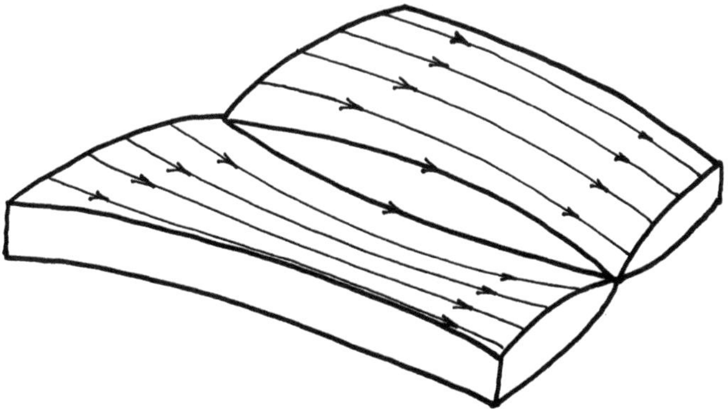

Fig. 10.3 A schematic drawing of the gap in a slab of magnetic field created by a local pressure maximum in the fields on either side of the gap.

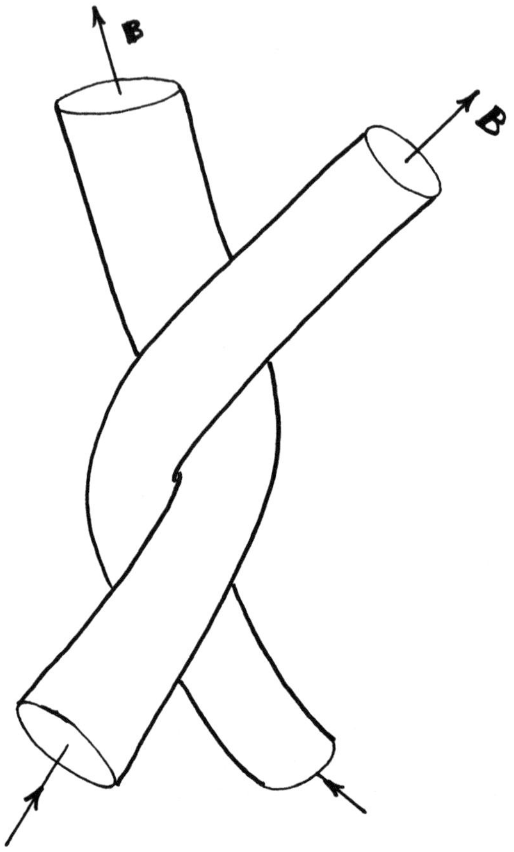

Fig.10.4 A schematic drawing of two flux bundles pulled partially around each other so that the tension in the bundles causes an increase in magnetic pressure where they are in contact.

would appear that if the field line interlacing extends through n distinct independent patterns between $z = 0$ and $z = L$, then each pattern forms its own TDs, unrelated to the TDs of patterns elsewhere along the field. Then, since TDs terminate only at the boundaries, this suggests that there may be as many as n TDs extending through the characteristic area of each successive pattern. The structure of the TDs is a complicated subject, discussed at greater length in Parker (1994, chapter 3).

The importance of the spontaneous formation of TDs lies in the rapid reconnection of magnetic field across each TD. For the Maxwell stresses, described by eqn. (10.1), drive the field toward the lowest available energy state (i.e., stable static equilibrium), with the TDs as the end result. That drive toward equilibrium is an asymptotic process, of course, presumably exponential, with a characteristic time of the order of L/C, and the final state involving TDs can be reached asymptotically only in the ideal case of the infinitely conducting fluid treated here.

In the real world there is always some slight resistivity, some slight friction as the electrons drift relative to the ions. So, the final singular equilibrium can be approached but never reached, and the Maxwell stresses continue to push the field toward sharper gradients and more concentrated current sheets. The system may achieve a quasi-steady state where the resistive dissipation softens the field gradient as rapidly as the Maxwell stresses sharpen it. This is the familiar *rapid reconnection* (of the field lines), because the dissipation and associated reconnection of the field across the thin current sheet at the incipient TD are driven dynamically by the Lorentz force, causing the dissipation to proceed very much faster than it would over the large-scale l of the overall field (Parker 1957b, 1963b; Sweet 1958).

10.6 Rapid Reconnection at an Incipient TD

Rapid reconnection is a dynamical process of relatively simple nature (Parker 1957b,1963b; Sweet 1958), sometimes not properly recognized. So we spell out the essential features here to be sure that there is no misunderstanding. Rapid reconnection occurs in those field configurations in which the Maxwell stresses push the field and fluid so as to steepen the field gradient (actually $\nabla \times \mathbf{B}$) without limit in the process of forming the ultimate TD essential for static equilibrium. The dynamics of the approach to a TD can be understood by examining what happens where two large-scale magnetic fields are pressed together across the surface $y = 0$, sketched in Fig. 10.5. The x axis lies in the $y = 0$ plane and is oriented so that one field makes an angle $+\vartheta$ with it, while the other field makes an angle $-\vartheta$. The plane of the figure is taken to be perpendicular to the x axis. The z axis lies in the plane of the figure, at right angles to the x axis, of course. Thus, in $y > 0$ we have $B_x = +B \cos \vartheta$, $B_z = +B \sin \vartheta$, while in $y < 0$ the field is given by $B_x = +B \cos \vartheta$, $B_z = -B \sin \vartheta$. So B_x is continuous across $y = 0$, but B_z is discontinuous.

Now, pressure balance along the y axis, as we pass from the field in $y < 0$ to the field in $y > 0$, requires that the total pressure of field and fluid be uniform and continuous. If $p(y)$ denotes the fluid pressure, this requirement along the y axis can be written

$$p(y) + \frac{B_x^2(y)}{8\pi} + \frac{B_z^2(y)}{8\pi} = P_0 \qquad (10.24)$$

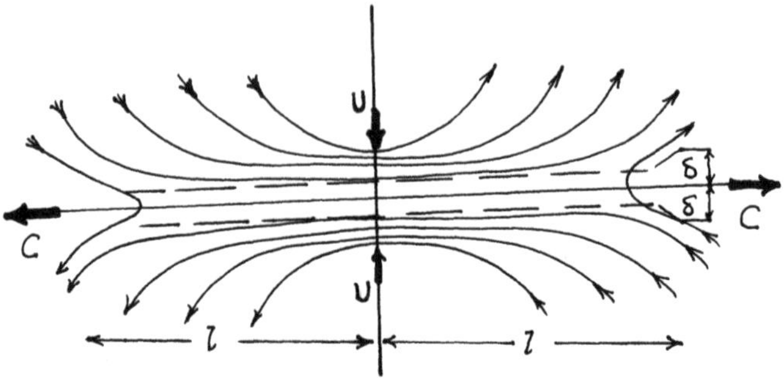

Fig. 10.5 A schematic drawing of the magnetic field in the *yz* plane where two nonparallel lobes of magnetic field press together across a front of width $2l$. The steep field gradient between the two opposite fields is confined to the thin layer of thickness 2δ. The resistivity dissipates the opposite fields in the steep field gradients, allowing the fields on either side to advance into the dissipation layer with the small speed u and expelling the fluid with the Alfvèn speed C out along the z axis

where P_0 is a constant. Now, in crossing $y = 0$, B_z changes from $+B \sin \vartheta$ to $-B \sin \vartheta$. In the ideal TD the jump is a true mathematical discontinuity. However, in the real world the jump is rapid but continuous. There is a thin transition region, with characteristic thickness denoted by 2δ, across which B_z declines from $+B \sin \vartheta$, passes through zero, and goes to $-B \sin \vartheta$. It is the fluid and field dynamics in that thin transition region that produces the rapid reconnection. Given that the total pressure P_0 is constant across the layer, the fluid pressure p increases by $B^2 \sin^2 \vartheta/8\pi$ at $y = 0$. So the fluid in that transition layer is expelled outward along the xz plane ($y = 0$). This causes the opposite field components to come closer together so that the field gradient increases while remaining subject to the same nonequilibrium stresses as before. The field gradient increases until finally resistive dissipation and diffusion become fast enough to prevent further increase. The field dissipation and reconnection then continues at a pace determined by the rate at which fluid can be expelled from between the two opposite fields.

With $B_z = 0$ on $y = 0$, it is clear that

$$p + \frac{B_x^2}{8\pi} = P_0 \tag{10.25}$$

instead of the condition

$$p + \frac{B_x^2}{8\pi} = P_0 - \frac{B^2 \sin^2 \vartheta}{8\pi}$$

that applies immediately on either side and at either end of the thin transition region. So the sum of the fluid and magnetic pressure at $y = 0$ is larger in the amount $B^2 \sin^2 \vartheta/8\pi$ than far away along the field lines where they are not squeezed between the two fields. So the fluid and the field B_x are squeezed out of the thin transition layer, pushed away in the $\pm z$ directions by the higher pressure.

The two large-scale field regions that press together at $y = 0$ have a scale which we denote by l, so that somewhere out there at distances of the order of l in the z direction the pressure falls to the ambient value. Neglecting viscosity the excess pressure does work on the ejected field and fluid in the amount $B^2 \sin^2 \vartheta/8\pi$ per unit volume, so that the kinetic energy of the ejected volume is given by

$$\tfrac{1}{2}NMv^2 = \frac{B^2 \sin^2 \vartheta}{8\pi} \tag{10.26}$$

It follows that the ejection velocity v is equal to the Alfvèn speed $C = B \sin \vartheta/(4\pi NM)^{1/2}$. It is this expulsion of fluid from the thin transition layer that causes the transition layer to become progressively thinner, enhancing the gradient between $B_z = +B \sin \vartheta$ in $y > 0$ and $B_z = -B \sin \vartheta$ in $y < 0$. This monotonic steepening of the gradient, and the associated concentration of the total electric current $cB \sin \vartheta/2\pi$ (per unit length in the x direction) into a progressively thinner layer, enhances the resistive dissipation up to the point where the rate of dissipation can keep up with the rate of steepening. In the simple case that the plasma is so dense that the simple scalar Ohm's law applies, the characteristic dissipation time τ across the thickness δ of the transition layer is $\delta^2/4\eta$, where η is the resistive diffusion coefficient $c^2/4\pi\sigma$. The characteristic speed at which the resistive dissipation can "eat" into the magnetic field on either side is of the order of $\delta/\tau = 4\eta/\delta$. Under steady conditions, then, the field on either side is driven by the Maxwell stresses into the dissipative transition layer with a speed $u = 4\eta/\delta$. The field can be pushed into each side of the transition layer only as fast as the fluid is pushed out along the field in the layer to escape from the open ends with the speed v. For an incompressible fluid the conservation of fluid requires that the inflow speed u over the breadth l is equal to the expulsion velocity v within the characteristic thickness δ, so that

$$ul = v\delta \tag{10.27}$$

in order of magnitude, where, we recall that $v = C$. Defining the Lundquist number

$$N_L \equiv \frac{Cl}{4\eta} \tag{10.28}$$

it follows (Parker 1957b) that

$$u = \frac{C}{N_L^{1/2}} \tag{10.29}$$

$$\delta = \frac{l}{N_L^{1/2}} \tag{10.30}$$

For comparison, the characteristic resistive diffusion velocity in the large-scale fields on either side of the concentrated transition layer is $4\eta/l$ or C/N_L. Thus, the magnetic dissipation and reconnection speed is enhanced by a factor of $N_L^{1/2}$ by the dynamical expulsion of the field and fluid. With $N_L \approx 10^7$–10^{22}, the enhancement is enormous (see discussion following eqn. (9.12)). Recent laboratory studies show this reconnection form and rate whenever the effective resistive dissipation is more or less uniform over the region (Yamada et al. 2000; Yamada 2001). However, observational studies find that the dissipation and reconnection are very much faster than $C/N_L^{1/2}$ in such phenomena as the solar flare, with u of the general order of 0.01–0.1C. This was baffling for several decades.

It has been pointed out in recent years that for the very large N_L to be found in nature, the thickness δ may be as small as the ion cyclotron radius and the ion inertial length. Thus, our MHD treatment of the dynamics of the dissipation layer is not valid within the current layer thickness δ. The transition, or dissipation, layer becomes much more complicated, requiring inclusion of the Hall effect and a full kinetic treatment of the ion and electron cyclotron motions. Fortunately, the modern computer with sophisticated programming is able to handle the problem, and it turns out that the dissipation rate is very much enhanced (Biskamp and Drake 1994; Drake et al.1994, 2005; Mandt et al. 1994; Biskamp et al. 1997; Shay and Drake 2001; Rogers et al. 2002; Jemella et al. 2003), bringing the theory into line with the rapid dissipation and reconnection inferred from observations of solar flares. It was discovered that when the kinetic effects take over from MHD, there is an additional enhancement of the dissipation and reconnection rate because the dissipation tends to concentrate in one or more localities rather than spread out uniformly along the length l of the transition layer. This localization may have a length $l' \ll l$, initiating a local reconnection scenario first proposed by Petschek (1964; Petschek and Thorne 1967). The idea is simply that the two opposite fields $\pm B \sin \vartheta$ may not press together over the large-scale l but rather touch only across a short distance l'. Petschek estimated that l' might be so small that the associated Lundquist number N'_L for the local reconnection region is only of the order of $\ln^2 N_L$. Thus, the reconnection rate becomes

$$u = \frac{C}{N'^{1/2}_L} = \frac{C}{\ln N_L} \qquad (10.31)$$

in order of magnitude. Evidently, this interesting configuration arises only when the dissipation mechanism is spotty, with the short dissipation layer centered on the concentrated dissipation. Laboratory experiments with rapid reconnection have begun to show how this works (Yamada et al. 2000; Yamada 2001; Ji et al. 2001; Yamada et al. 2006).

Besides the kinetic effects, it should be appreciated that the MHD steady-state reconnection for which eqn. (10.29) was developed is an idealization. In the real MHD world resistive instabilities are likely to develop, breaking up the transition layer into local magnetic islands around O-type neutral points with X-type neutral points where neighboring islands meet. Thus, transient reconnection may occur simultaneously at several places along the transition layer (Biskamp 1986; Wang and Bhattacharjee 1992). This alone may cause a modest enhancement of the overall reconnection rate, unless, of course, it initiates the Hall and kinetic effects that provide the really rapid reconnection already mentioned (see Parker 1994, section 10.3.5 for a detailed discussion and references).

Now, in many circumstances in the laboratory and in astrophysics, the rapid reconnection proceeds at the slow rate indicated by eqn. (10.29). In the evolving magnetic fields of active regions observed on the Sun, for instance, the reconnection may require a day or a month, while the occasional break over into kinetic dissipation appears to be responsible for flares with onset times of 10^2 s. and total magnetic dissipation ranging from 10^{32} ergs down to somewhere below the present observational limit of about 10^{24} ergs. So the physics of rapid reconnection has come a long way in the half-century since it was first proposed. The complexity of the plasma instabilities and the kinetic effects that so greatly enhance the basic rate of eqn. (10.29) make it clear that one does not calculate with any precision the reconnection rate in most circumstances in nature.

The general point is that the large-scale magnetic fields in the cosmos are embedded in the swirling ambient plasma, thereby developing TDs and rapid reconnection in the interlacing components of the field. The dynamics drives the thickness of the incipient TD down to such small scales that there is rapid dissipation of the interlacing, or transverse components, no matter how slight the effective resistivity might be. Indeed, the larger the Lundquist number N_L, the more likely it is that δ falls to the ion cyclotron radius, bringing on the non-MHD plasma kinetics that provide such explosive dissipation of magnetic energy. Thus, the large-scale magnetic fields tend to dissipate any "untidy" aspects of their internal topology, the dissipated magnetic energy heating the tenuous local plasma.

Note, finally, that the dissipative current sheet in the plane $y = 0$ in the dynamical rapid reconnection process possesses a significant electric field $\mathrm{E}_\|$ parallel to the mean magnetic field $B_x = B \cos \vartheta$ in the x direction. It is this electric field that drives the intense dissipative electric current in the transition layer $-\delta < y < +\delta$. For steady reconnection, $\partial \mathbf{B}/\partial t = 0$ it follows from eqn. (1.5) that $\nabla \times \mathbf{E} = 0$, so $\mathrm{E}_\|$ is approximately uniform across the thickness of the current sheet. It is related to the inflow velocity in the y direction by $E_\| = E_x = uB \sin \vartheta / c$. It has been pointed out that $E_\|$ may accelerate ions and electrons along the mean magnetic field, accounting for some of the fast particle production in solar flares and in the terrestrial aurora. The reader may be interested in the paper by Schindler et al. (1991), where they treat the relation of $E_\|$ to the absence of isorotation in magnetic fields with rotational symmetry about the z axis.

10.7 Quasi-steady Dissipation at a TD

It is evident that the rapid reconnection process dissipates the transverse interlaced components of the field lines. The field configuration described by eqn. (13.3) evolves back toward the original, uniform, field B in the z direction. In nature, of course, the convective swirling of the footpoints of the field at $z = L$ never ceases, so the rapid dynamically driven dissipation goes on and on. Figure 10.6 is a sketch of the interlacing within a bipolar magnetic field of an active region on the Sun, with both ends rooted in the convective surface. One expects that there is continuing strong dissipation of the magnetic energy of the interlacing component of the magnetic field, and this seems to be the major heat source responsible for the X-ray coronal loops observed on the Sun.

The trend toward TDs in both stellar and galactic magnetic fields is universal, suggesting a continuing conversion of magnetic energy into heat, at rates equal to the rate at which work is done by the convective swirling of the footpoints of the field. The simple model of one end of a bipolar magnetic field sketched in Fig. 10.7, ignores the curvature of the magnetic loop of large length L, and illustrates how the random walk, at speed v, of the footpoint of a single flux bundle in an ambient vertical uniform field B does work on the field. The random motion of the footpoint of the bundle through the ambient forest of vertical flux bundles

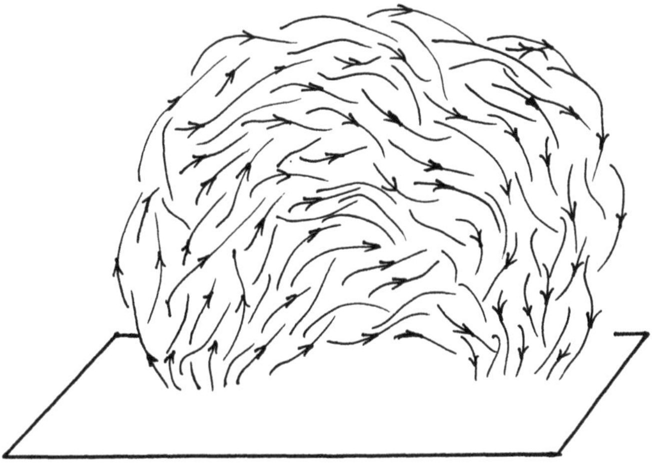

Fig. 10.6 A schematic drawing of the interlaced field lines of a bipolar magnetic field, typical of an active region on the Sun, where the photospheric footpoints of the field are continually swirled and intermixed by the subsurface convection.

has a total travel distance vt after a time t. This causes the flux bundle to be inclined by an angle θ relative to the vertical, where

$$\tan\theta = \frac{vt}{L}$$

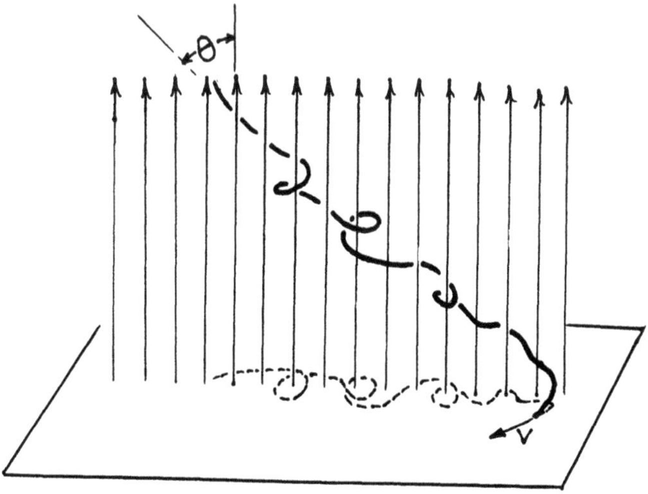

Fig. 10.7 A schematic drawing of the field lines of the vertical ambient field issuing from an active region on the Sun, showing slender flux tube (heavy line) whose photospheric footpoint random walks with speed v. The path of the footpoint is indicated by the dashed line.

The vertical field component B remains more or less the same and the horizontal component B_n is

$$B_n = B \tan \vartheta$$
$$= \frac{Bvt}{L}$$

The Maxwell stress pulling back against the forward-moving footpoint is $BB_n/4\pi$ per unit area, so that the forward motion v does work on the field at the rate

$$W = v\frac{BB_n}{4\pi}$$
$$= \frac{B^2v^2t \text{ ergs}}{4\pi L \text{ s cm}^2}$$

Under steady conditions the energy dissipation consumes this energy as rapidly as it is introduced, so W is the effective dissipative heat input to the ambient gases in the flux bundle.

To use an X-ray loop on the Sun as an example, put $W = 10^7$ergs/cm^{2}s with $B = 10^2$G and $v = 1$ km/s. It follows that

$$\frac{vt}{L} = \frac{4\pi W}{B^2 v}$$
$$\cong 0.1$$

That is to say, $\tan \vartheta \sim 0.1$ and ϑ is of the order of 6^0. So the interlacing of the field does not have to be very strong to supply the heat input inferred from the observed radiative (X-ray) cooling (Parker 1981a, b, 1983, 1988, 1994).

There is direct observational evidence of such heating in the X-ray corona of the Sun. Low and Berger (2002) show that the observed sigmoid structures in the X-ray emitting corona can be understood in terms of spontaneous surfaces of tangential discontinuity. Then recall that both laboratory and numerical experiments show that the magnetic dissipation occurs in numerous small bursts of reconnection, or *nanoflares* (Parker 1988). It is interesting to note, then, that theoretical analysis of the observed coronal emission spectrum shows that the spectrum cannot be understood in terms of a time-independent temperature, even if a number of different steady temperatures are combined. The emission spectrum can be understood only as a result of episodic heating, in which the electron temperature frequently jumps to values of the order of 10^6–10^7 K and then relaxes over a period of minutes. The ionization states of the atoms respond relatively slowly to the individual electron temperature transients, so that part of the time the electron temperature differs from the equivalent ionization temperature (Sturrock et al. 1990; Feldman et al. 2005). More recently Katsukawa (2003; Katsukawa and Tsuneta 2005) have studied the fluctuations in the individual pixels of the Yohkoh X-ray telescope. They find fluctuations in excess of the thermal background, indicating large numbers of small-scale flares with individual events in the range 10^{21}–10^{24} ergs—picoflares and nanoflares. Klimchuk (2006) provides a critical look at the feasibility of the diverse ideas that have been proposed for heating the solar X-ray corona.

These analyses support the concept that the X-ray corona is created by a large number of very small flares in the TDs throughout the interlaced bipolar magnetic fields of active regions, all driven by the granule convective cells at the photosphere. If this is correct, then it may be inferred that most other ordinary stars produce their X-ray emission in the same way. That is to say, we have universal stellar X-ray emission as a consequence of the remarkable singular property of the Maxwell stress tensor in a bipolar magnetic field with interlaced field lines.

Needless to say, the Galaxy and the galactic halo present a somewhat different picture. The principal energy input to the magnetic field of the galactic halo appears to be the 30- to 50-km/s outward inflation of the extended Ω-loops of the galactic magnetic field by the cosmic ray gas pumped into the field by the supernovae in the gaseous disk of the Galaxy (Parker 1992). The rapid reconnection between adjacent lobes of inflated magnetic field appears to be a major source of heat to the tenuous halo gas, at X-ray emitting temperatures of 10^6– 10^7 K.

Now, we must be aware that the simple theoretical analysis provided here is intended only to establish the plausibility of heating the stellar X-ray corona by the relaxation of the untidy magnetic field topology to form surfaces of tangential discontinuity. The actual situation on the Sun involves some additional modulating factors of which we should be aware. Note, then, that the magnetic field of the corona extends upward from the photosphere, where the electrical resistivity is relatively high ($\sigma \approx 10^{11}$/s, $\eta \approx 10^9$ cm^2/s), through the chromosphere and the transition region into the very small resistivity ($\sigma \approx 10^{16}$/s, $\eta \approx 10^4$ cm^2/s) of the million degree corona. Another point is that the continual convective intermixing of the photospheric footpoints of the magnetic field might be expected to create steep transverse field gradients independently of the spontaneous steepening of the gradients in the field in the corona. At first sight it might appear that the photospheric dissipation, where the resistivity is 10^5 times larger than in the corona, would swamp the prospect for steep field gradients in the corona. However, it must be remembered that the magnetic field in the photosphere is confined to widely separated magnetic fibrils with characteristic diameters of the order of 10^7cm and field strengths of $1–2 \times 10^3$ G. In an active region, where the mean field is of the order of 10^2 G, the fibrils occupy only 0.05–0.10 of the total area. So the individual fibrils interact with each other only occasionally, when one bumps into one of its neighbors. Microflaring may occur at such times if the colliding fibrils have opposite fields. It would appear that such interactions temporarily and locally modulate and possibly disrupt the formation of tangential discontinuities in the corona above, providing a complicated photosphere–corona interaction.

The fibrils, extending upward through the chromosphere and transition region, expand with height and begin to press against each other at elevations of 10^3km or more. The continual convective shuffling of the photospheric fibrils suggests that the fields of two contiguous fibrils are not exactly parallel where their fields press together in the transition region and low corona, providing a steep field gradient in addition to the tangential discontinuities formed by the overall untidy topology. The characteristic steepening time associated with the photospheric convective cell (granule) with velocity $v \approx 1$ km/s over a characteristic scale $l \approx 500$ km is $l/v = 500$ s. In the corona the characteristic steepening time L/C may be as large as 50 s, where $L \approx 10^5$ km for the total length of coronal loop and the Alfvèn speed C is 2×10^3km/s for a magnetic field of 100 G and a plasma density of 10^{10} ions/cm^3. So the steepening proceeds more rapidly in the corona, but not so rapidly that conditions in the photosphere, chromosphere, and transition region cannot sometimes affect the rate of steepening of the field gradients in the corona. That is to say, coronal reconnection and dissipation of magnetic energy may be modulated in interesting ways by conditions in the chromosphere and transition region and by the height at which the fields of neighboring fibrils begin to press against each other. This complicated scenario has not yet been investigated, and it will need the observational guidance of the highest possible telescopic resolution of the fibril structures at the Sun.

11 Comments

11.1 Summary

It is time to wind down the conversation with a number of general remarks. To rehash the foregoing conversation, recall that we began with a review of the concepts of electric and magnetic phenomena in the cosmos. The cosmos is filled with magnetic fields and the magnetic field is filled with plasma to which it is strongly coupled. Thus, what began as a study of electromagnetic theory became an exercise in hydrodynamics (HD) and magnetohydrodynamics (MHD). The electric fields, which are unique and isolated in each different moving reference frame, drop out of the dynamics along with the electric current. The Faraday induction equation for the large-scale magnetic field **B** becomes (eqn. (9.2))

$$\frac{\partial \mathbf{B}}{\partial t} = \nabla \times (\mathbf{v} \times \mathbf{B})$$

because there is no significant electric field in the frame of reference of the plasma moving with large-scale bulk velocity **v**. The plasma velocity **v** satisfies the momentum equation (9.5), which includes the force $\partial M_{ij}/\partial x_j$ exerted by the magnetic field on the plasma or fluid. The magnetic field is transported and deformed by the swirling plasma velocity **v**. So the plasma inertia and pressure become an intimate part of the electromagnetic field dynamics. The dynamics, then, is an exercise in the contending forces of **v** and **B**. The electric current can be calculated from Ampere's law, eqn. (3.2), if it is needed for the resistive dissipation of magnetic energy and the heat input to the plasma. We provided a discussion of the various dissipative terms that may appear on the right-hand side of eqn. (9.2). In large-scale fields those dissipation terms, as well as the nondissipative Hall term, are small compared to the large-scale term $\nabla \times (\mathbf{v} \times \mathbf{B})$. Exceptions are important, of course, in the small scales of concentrated current sheets, boundary layers, and shock fronts, which give rise to such exotic phenomena as magnetic flares and planetary aurora.

The electric field can be calculated from $-\mathbf{v} \times \mathbf{B}/c$ in whatever moving frame of reference is of interest, and it must be appreciated that each moving frame is its own little world. The relatively strong electric field in a rapidly moving frame does not reach over to drive the plasma in a slower moving frame, as sometimes imagined for the active terrestrial magnetosphere and for the acceleration of the vast populations of fast particles to be found around active celestial objects.

11.2 Electric Circuit Analogy

Now, no conversation on the magnetic fields in the cosmos would be complete without a comment on the fantasies that have replaced physics in so many applications. So a presentation of that sociological subject forms the coda of this conversation. The underlying problem seems to be a lack of appreciation of the fact that physics is constrained by the basics laws of nature. In the present case, those laws are contained in the equations of Newton and Maxwell, and Lorentz and Boltzmann. An assertion about magnetic fields and plasmas is not physics unless it can be shown to follow from Newton and Maxwell et al., and scientific calamity arises when anyone strays from that principle. A striking example, among many, has been presented recently by Parks (2004), who fails to distinguish the large-scale domain of MHD from the small-scale plasma kinetics of thin boundary layers and current sheets. Most striking of all is his assertion that since $\mathbf{E}'$ is *identically* zero in the ideal infinitely conducting fluid, it follows that "no currents and magnetic fields are generated in the fluid. Ideal MHD theory removes the capability for the plasma to act electromagnetically. This restriction severely limits the kind of physics one can do with ideal fluids." Parks fails to recognize that ideal MHD occurs in the limit of diminishing $\mathbf{E}'$ as the electrical sensitivity of the plasma increases without bound. The reader is urged to consult this paper first hand to get some appreciation of how far afield the subject has wandered.

The intellectual derailment appears to begin with the idea that magnetic fields are *caused* by electric currents, as indeed they are in the laboratory when we apply a voltage to a coil of wire to produce a magnetic field. The magnetic field in the laboratory is intimately linked to **E** and **j**, sometimes called the **E**, **j** paradigm (cf. Parker 1996a; Vasyliunas 2001, 2005a,b; Vasyliunas and Song 2005). Now, we all learned electric circuit theory early in our careers. This seems to make it difficult to accept the idea that, in a large-scale swirling plasma, it is the magnetic field that drives the current, through the $\partial \mathbf{E}/\partial t$ in Maxwell's equation (6.7) and through the Newtonian mechanics of the ion and electron trajectories in a collisionless plasma. The energy to drive the currents comes from the magnetic field. So the field *causes* the current. The current is secondary, an effect of the magnetic field rather than vice versa. It is the interaction of **B** and **v** that dominates the dynamics—the **B**, **v** paradigm. It should not be surprising, then, that we cannot formulate a complete set of workable field equations in terms of **E**, **j** from the principles of Newton and Maxwell. To do so leads to an intractable set of global nonlinear integro-differential equations, wherein the conditions at any given point in space are expressed in terms of integrals of $\mathbf{j}(\mathbf{r}, t)$ over the entire system. Faced with this fundamental road block, the practice of **E**, **j** sometimes drifts into fantasy, substantiated by the authority of widespread consensus. As already noted, the electric field $\mathbf{E} = -\mathrm{v} \times \mathbf{B}/c$ in the laboratory frame is assigned a dynamical role in modifying the motion of the plasma.

The diverse misconceptions are epitomized by the belief that the dynamics of a time-dependent magnetic plasma configuration can be represented by an equivalent simple electric circuit. This concept is seductive and is the basis for many "scientific" papers in the journals. Unfortunately, equivalent laboratory circuits for time-dependent dynamical magnetic plasma systems contradict the MHD properties of the plasma deduced from Newton and Maxwell.

For instance, the electric circuit in the laboratory consists of conducting channels (wires) that are fixed in the frame of reference of the laboratory. The topology and connectivity of the channels are fixed and do not change during the operation of the circuit. On the other hand, the topology of the electric current in the swirling plasma is not fixed. The magnetic field is swirled along with the plasma, and the current is determined by Ampere's law and the changing topology of the swirling $\nabla \times \mathbf{B}$. Equally important, the current in the plasma flows in the frame of reference of the moving plasma, and in that frame of reference there is no significant electric field. The laboratory circuit, on the other hand, is fastened to the workbench and cannot avoid the electric fields in the frame of reference of the laboratory. So the current in the plasma does not experience the effects of inductance that may have such powerful effects in the fixed laboratory circuit.

Then it must be appreciated that variations at one location in the laboratory electric circuit are communicated to the rest of the circuit at, more or less, the speed of light. So the electrodynamics is tightly coupled across the entire circuit. In contrast, communication across a large-scale magnetic field in a plasma is restricted to the lumbering Alfvèn speed, comparable in magnitude to the plasma velocity. The dynamics at one end of the system is not felt at the other end until some time later. Magnetic energy converts into plasma motion only with the passage of an MHD wave, so the conversion is progressive rather than explosive.

11.3 A Simple Example of an Electric Circuit

It is instructive to venture from these generalities to a simple illustrative example to show how the plasma–field system responds to something as drastic as, say, a sudden blocking of the flow of electric current. Now the electric circuit analog asserts that the inductance L of a current loop in a moving plasma is defined by equating the magnetic energy associated with the loop to $\frac{1}{2}LI^2$, where I is the total electric current flowing around the loop. The belief is that, if the current is abruptly impeded in some way, e.g, the onset of plasma turbulence and anomalous resistivity, or an electric double layer, then a large electric potential difference LdI / dt arises across the region of obstruction (cf. Alfvèn and Carlquist 1967). On this basis an abrupt blockage of I would result in accelerating some of the local free charges to large energies. The creation of energetic auroral particles, fast particles from solar flares, and even galactic cosmic rays is suggested.

What happens in a plasma is quite different from the prediction of the circuit analog. First of all, placing an obstacle in the current path in a plasma results in the current immediately finding a new path around the obstacle. Thus, there is no immediate change in the current flow elsewhere in the system. That is to say, the magnetic field quickly restructures itself locally so that Ampere's law provides a current that bypasses the obstacle. Then the restructuring of **B** upsets the local balance of magnetic stresses, so that the plasma is set in motion in such a way as to avoid any significant electric field in its own frame of reference. We have already noted that $\mathbf{E}' = 0$ and $\mathbf{E} = -\mathbf{v} \times \mathbf{B}/c$ represent a self-consistent premise in an infinitely conducting fluid. Newton's equations of motion and the magnetic stresses consistently maintain that condition.

To work out a specific case, consider a long, twisted flux bundle extending along the z axis and confined within a radius R of the z axis by the external pressure of the surrounding uniform (density ρ), infinitely conducting, incompressible fluid that pervades the entire space. There is an electric current flowing within the twisted flux bundle. In the simplest case the current flows one way along the central core of the bundle and the opposite way near the periphery of the bundle. The total current is zero, of course, which follows from the integral form of Ampere's law, eqn. (3.12), applied to a path circling the flux bundle beyond R.

Denoting the azimuthal and longitudinal components of the magnetic field by $B_\varphi(\varpi)$, $B_z(\varpi)$, respectively, where ϖ represents radial distance $(x^2 + y^2)^{1/2}$ from the z axis, it follows from Ampere's law that the current density is

$$j_\varphi(\varpi) = -\frac{c}{4\pi}\frac{dB_z}{d\varpi}$$

$$j_z(\varpi) = +\frac{c}{4\pi}\frac{1}{\varpi}\frac{d}{d\varpi}(\varpi B_\varphi)$$

To fix ideas, suppose that $B_\varphi(\varpi) \geq 0$. It is easy to show that $j_z(\varpi) = 0$ at the radius $\varpi = a$ where B_φ varies inversely with ϖ, i.e., where

$$\frac{1}{B_\varphi}\frac{dB_\varphi}{d\varpi} \approx -\frac{1}{\varpi}$$

For simplicity, assume that there is only one such radius, and suppose that B_z is a maximum and $B_\varphi = 0$ on the z axis ($\varpi = 0$). It follows that $j_z(\varpi) \geq 0$ over the central region $0 < \varpi < a$, and $j_z(\varpi) \leq 0$ over the outer region $a < \varpi < R$. The total current I flowing upward along the z axis is

$$I = 2\pi \int_0^a d\varpi \varpi j_z(\varpi)$$

$$= \tfrac{1}{2} c a B_\varphi(a)$$

The total current flowing downward beyond the radius $\varpi = a$ has the same magnitude, of course. The magnetic energy W per unit length of flux bundle associated with the current I is

$$W = 2\pi \int_0^R d\varpi \varpi \frac{B_\varphi^2(\varpi)}{8\pi}$$

The inductance L per unit length is given by the relation

$$\tfrac{1}{2} L I^2 = W$$

so that

$$L = \frac{1}{c^2 a^2 B_\varphi^2(a)} \int_0^R d\varpi \varpi B_\varphi^2(\varpi)$$

It follows that the total inductance in the current path over a total length 2Λ is $2\Lambda L$, and can be made arbitrarily large by making the flux bundle arbitrarily long.

At time $t = 0$ a slab of electrical insulating material of thickness 2ε instantaneously slices across the flux bundle at $z = 0$, blocking the passage of electric current. According to the electric circuit analog, the large inductance of the long flux bundle produces an immense potential difference across the slab of insulation at the moment the current is blocked. However, a careful examination of the physics determines otherwise.

Figure 11.1a is a sketch of the magnetic field configuration at the instant the slab first appears across the flux bundle. The azimuthal magnetic field $B_\varphi(\varpi)$ in $-\varepsilon < z + \varepsilon$ suddenly finds itself in a nonconducting material with $j_z = 0$. It follows that the Maxwell relation

$$\frac{\partial E_z}{\partial t} = +\frac{c}{\varpi}\frac{\partial}{\partial \varpi}(\varpi B_\varphi)$$

replaces Ampere's law. Then, with the induction equation,

$$\frac{\partial B_\varphi}{\partial t} = +c\frac{\partial E_z}{\partial \varpi}$$

the azimuthal component of the magnetic field suddenly finds itself to be an electromagnetic wave, propagating away at the speed of light (in the insulating material of the slab).

With the escape of B_φ from $-\varepsilon < z < +\varepsilon$ the magnetic field has the form sketched in Fig. 11.1b. The kinks in the magnetic field at $z = \pm\varepsilon$ provide a Lorentz force accelerating the fluid in the φ direction and initiating torsional Alfvèn waves propagating away along the flux bundle with the Alfvèn speed $C(\varpi) = B_z(\varpi)/(4\pi\rho)^{1/2}$ at the radial distance ϖ. The wave initiated at $z = +\varepsilon$ propagates away in the $+z$ direction; while the wave from $z = -\varepsilon$ propagates in the $-z$ direction; sketched in Fig. 11.1c. The passage of the torsional wave front reduces the initial $B_\varphi(\varpi)$ to zero, setting the fluid behind the wave front in azimuthal motion with velocity $v_\varphi(\varpi) = B_\varphi(\varpi)/(4\pi\rho)^{1/2}$. Note that the radial form of the wave changes with time because B_z is a function of ϖ. The dynamics is developed in appendix F for the interested reader, where it is shown that the force exerted by the magnetic field on the fluid at the kink is just sufficient to produce the stated azimuthal velocity $v_\varphi(\varpi)$. Note, however, that conservation of energy,

$$\tfrac{1}{2}\rho v_\varphi^{2}(\varpi) = \frac{B_\varphi^{2}(\varpi)}{8\pi}$$

describes the conversion of B_φ into the rotational motion v_φ more directly. The essential point is that the magnetic energy with which the blocked electric current is associated is converted into the motion of an Alfvèn wave rather than providing a large emf, which might have accelerated particles to very high speeds.

This simple example illustrates the inadequacy of the electric circuit analog for time-dependent systems. One cannot handle dynamical problems with the circuit analog, nor indeed, is the (**E**, **j**) paradigm itself generally

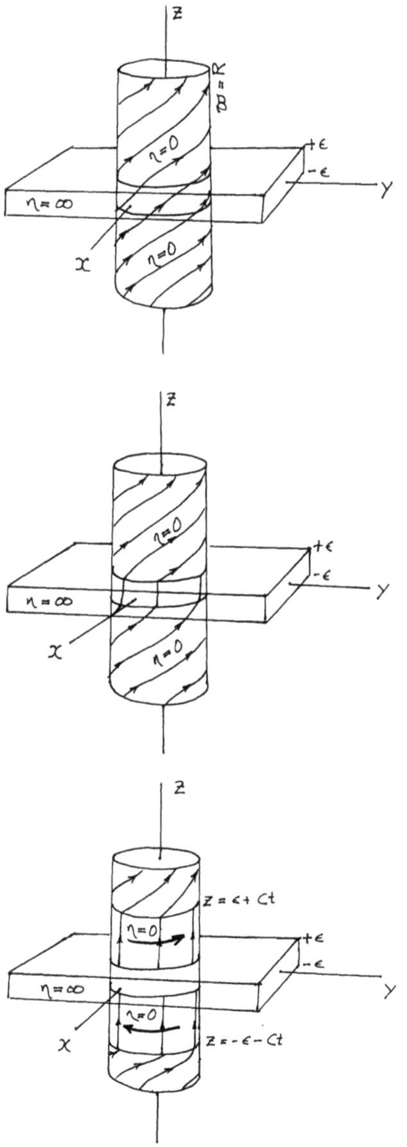

Fig. 11.1 (a) A schematic drawing of the twisted flux bundle with its initial configuration at the instant the insulating slab $(-\varepsilon < z < +\varepsilon)$ is inserted. (b) The field configuration immediately following the escape of B_φ from the nonconducting region $(-\varepsilon < z < +\varepsilon)$. (c) A schematic drawing of the magnetic configuration and fluid motion a time t later showing the two torsional Alfvèn waves propagating away in opposite directions along the twisted flux bundle, leaving $B_\varphi = 0$ behind the wave front and setting the plasma and field in rotation with azimuthal velocity $v_\varphi = \pm B_\varphi/(4\pi\rho)^{1/2}$.

able to handle the large-scale dynamics of a time-dependent magnetic field and plasma system. It is to be hoped that the continuing dynamical futility of the (**E**, **j**) paradigm, and the calamitous ideas it has spawned, can be replaced by direct attention to the physics of Newton and Maxwell.

11.4 Popular Electric and Magnetic Fields

To end the conversation on a lighter note, the widespread appeal of electric and magnetic fields to the general public is to be noted. Last year my wife and I spent a few days hiking in the desert and mountains around Sedona, Arizona, and we were interested to learn of the local interest in electric and magnetic fields. Perhaps the best description of the public interest is to quote the local authorities (Samson and Aukshunas, 2002):

> In recent years, Sedona has been one of the world's centers for the New Age movement; large numbers of people come to experience the "power vortices" of the surrounding red rock country . . . According to believers, a vortex is a site where the earth's unseen lines of power intersect to form a particularly powerful energy field. Page Bryant, an adherent to New Age beliefs, determined through channeling that there are four vortices around Sedona. . . . The four main vortices include Bell Rock, Cathedral Rock, Airport Mesa, and Boynton Canyon. Bell Rock and Airport Mesa are both said to contain masculine or electric energy that boosts emotional, spiritual, and physical energy. Cathedral Rock is said to contain feminine or magnetic energy good for facilitating relaxation. . . . The Boynton Canyon vortex is considered an electromagnetic energy site, which means it is a balance of both masculine and feminine energy.

Then,

> Some people believe that if the vortex is too strong, it can eventually weaken the immune system. Others have even reported that compasses and electronic devices, such as cell phones and watches, don't work properly around these energy fields.
>
> *Red Rock Country Visitors Guide*, 2004

Our hiking took us around Bell Rock and up Boynton Canyon, and then a drive up Airport Mesa for the superb view of the surrounding mountains and desert. Evidently, the vortices were switched off those days because our compass and watches did not falter, and our immune systems emerged unscathed. Aircraft operating in and out of the Sedona airport reported no anomalies with either their radio communication or their direction-finding electronics. But it shows the fascination that electric and magnetic fields hold for many individuals, adding a new dimension to electricity and magnetism and illustrating once again the intellectual calamities that arise when one departs from Newton and Maxwell.

Appendix A Electrostatically Driven Expansion of the Universe

If electrostatic repulsion is to account for the expansion of the universe, that repulsion must overpower the gravitational attraction. The positive electrostatic potential of a spherical volume of radius R must exceed the magnitude of the negative gravitational potential. The gravitational potential energy V_G of a sphere of radius R of uniform number density N of baryons, each of mass M, is

$$V_G = -\frac{3GK^2}{5R}$$

where K is the total mass,

$$K = \tfrac{4}{3}\pi R^3 NM$$

The electrostatic potential energy of the same sphere, in which each electron–proton pair has a net charge $\pm\varepsilon e$, is

$$V_E = +\frac{3Q^2}{5R}$$

where Q is the total charge,

$$Q = \tfrac{4}{3}\pi R^3 N\varepsilon e$$

Hence, electrostatic expansion requires that

$$\varepsilon e > G^{1/2}M$$

With M equal to the mass 1.67×10^{-24}g of the hydrogen atom, $e = 4.8 \times 10^{-10}$ esu, and $G = 6.667 \times 10^{-8}$ cm^3/gs^2, the condition becomes $\varepsilon > 0.9 \times 10^{-18}$. Thus, in one mole ($6 \times 10^{23}$ particles) there must be the equivalent of more than 5.4×10^4 extra electronic charges, or 2.6×10^{-5}esu. Observational estimates of the Hubble constant and the mean density of the universe suggest that the present kinetic energy of the expansion is nearly equal in magnitude to the negative gravitational potential energy ($\Omega \approx 1$). This implies that the electrostatic repulsion represents a potential energy of twice the gravitational potential energy, in which case $\varepsilon \approx 1.8 \times 10^{-18}$, and a mole of matter has a charge of 5.2×10^{-4}esu.

A brief look at the time-dependent dynamics is not without interest. A uniform radial expansion throughout the sphere with the surface $r = R$ is described by the velocity profile$(d \ln R/dt)r$ throughout the sphere. So the total kinetic energy T in $r < R$ is given by

$$T = \frac{3}{10}K\left(\frac{dR}{dt}\right)^2$$

The total energy U of the system is constant and $U = T + V_E + V_G$. Thus,

$$U = \frac{3}{10}K\left(\frac{dR}{dt}\right)^2 + \frac{3(Q^2 - GK^2)}{5R}$$

If the expansion started from rest from some small radius a, then

$$U = \frac{3(Q^2 - GK^2)}{5a}$$

It follows that the subsequent expansion velocity dR/dt is given by

$$\frac{dR}{dt} = \left[\frac{2(Q^2 - GK^2)}{aK}\right]^{1/2}\left(1 - \frac{a}{R}\right)^{1/2}$$

It is evident, then, that by the time R is large compared to the initial small radius a, the expansion has approached very closely to the final asymptotic value,

$$\frac{dR}{dt} = \left[\frac{2(Q^2 - GK^2)}{aK}\right]^{1/2}$$

That is to say, the acceleration is concentrated in the early moments of the expansion and plays no role in the acceleration of expansion inferred from present-day observations of the expansion.

Appendix B Relaxation of Electric Charge Inhomogeneity

Consider what happens when a small number density $n_0(\mathbf{r})$ of electrons is introduced into a cold, uniform, electrically neutral plasma composed of N electrons and N protons per unit volume. Denote the proton mass by M and the electron mass by m, with $m << M$, of course. There is, then, a nonvanishing charge density $\delta(\mathbf{r}) = -en_0(\mathbf{r})$ and an associated electric field $\mathbf{E}$, given by

$$\nabla \cdot \mathbf{E} = -4\pi en(\mathbf{r}, t) \tag{B1}$$

where the density $n(\mathbf{r}, 0)$ has the initial value $n_0(\mathbf{r})$. The electrons start from rest with a velocity $\mathbf{u}$, described by

$$m\frac{d\mathbf{u}}{dt} = -e\mathbf{E} - \frac{m\mathbf{u}}{\tau} \tag{B2}$$

The last term represents friction through collisions with the massive and essentially immobile ions, with the effective mean collision time τ. Assuming that $n_0(\mathbf{r})<<N$, so that the electron velocity $\mathbf{u}$ is small compared the characteristic scale of $n_0(\mathbf{r})$ multiplied by the electron plasma frequency, the equation of motion can be linearized and $d\mathbf{u}/dt$ replaced by $\partial\mathbf{u}/\partial t$, with

$$\frac{\partial\mathbf{u}}{\partial t} + \frac{\mathbf{u}}{\tau} = -\frac{e\mathbf{E}}{m} \tag{B3}$$

Conservation of electrons is described by the linearized equation

$$\frac{\partial n}{\partial t} + N\nabla\cdot\mathbf{u} = 0 \tag{B4}$$

The divergence of (B3) can then be written in terms of $n(\mathbf{r}, t)$ upon using eqns. (B1) and (B4), yielding

$$\frac{\partial^2 n}{\partial t^2} + \frac{1}{\tau}\frac{\partial n}{\partial t} + \omega_{\mathrm{p}}{}^2 n = 0 \tag{B5}$$

where $\omega_p = (4\pi Ne^2/m)^{1/2}$ represents the electron plasma frequency. We are interested in solutions of the form $n(\mathbf{r}, t) = n_0(r)\exp i\omega t$, for which

$$\omega = \pm\omega_p\left(1 - \frac{1}{4\omega_p^2\tau^2}\right)^{1/2} + \frac{i}{2\tau} \tag{B6}$$

It is evident by inspection that when $\omega_p\tau > \frac{1}{2}$, the oscillations decay as $\exp(-t/2\tau)$. When $\omega_p\tau < \frac{1}{2}$, it follows that

$$i\omega\tau = -\frac{1}{2}\left[1 \pm (1 - 4\omega_p^2\tau^2)^{1/2}\right]$$

The quantity in square brackets is positive, so the system relaxes exponentially to rest, with the lower sign giving the slowest decay. As $\omega_p\tau$ becomes small compared to one with increasing N,

$$i\omega \approx -\omega_p^2\tau$$

Noting that the collision time τ is equal to the mean free path λ divided by the electron thermal velocity, $w_e=(kT_e/m)^{1/2}$, write

$$\lambda = \frac{1}{NA(T_e)}$$

where $A(T_e)$ is the effective Coulomb collision cross section (cf. Spitzer 1956), and T_e is the effective electron temperature. Thus,

$$\tau = \frac{1}{NA(T_e)}\left(\frac{m}{kT_e}\right)^{1/2}$$

and

$$\omega_p^2\tau = \frac{4\pi e^2}{A(T_e)(mkT_e)^{1/2}}$$

Note, then, except for a logarithmic variation with N in $A(T_e)$, this damping rate is independent of the density in the limit of large N.

In the strong collision limit it is convenient to write $\mathbf{j}=\sigma\mathbf{E}$, where σ is the uniform electrical conductivity. Then with $\nabla\cdot\mathbf{E}=4\pi\delta$, the condition $\nabla\cdot\mathbf{j} + \partial\delta/\partial t = 0$ for conservation of electric charge reduces to $4\pi\sigma\delta + \partial\delta/\partial t = 0$, so that $\delta(t) = \delta(0)\exp(-4\pi\sigma t)$.

In the realistic case that the plasma is hot rather than cold, Landau damping dominates the picture for all wave numbers k such that $k\lambda_D > 1$, where the Debye radius is

$$\lambda_D = \left(\frac{kT_e}{4\pi Ne^2}\right)^{1/2}$$

$$= \frac{w_e}{\omega_p}$$

That is to say, the Debye length is the characteristic distance traveled by a thermal electron in the time $1/\omega_p$. When that distance becomes comparable to the wavelength of the plasma oscillation the electrons begin to escape from the wave and the wave or oscillation loses coherence. Numerically, $\omega_p \cong 5.6 \times 10^4 N^{1/2}$ s. The damping of the perturbation $n_0(\mathbf{r})$ is thus more rapid than by Coulomb collisions alone.

Note that the electron response to a charge perturbation is so fast that the ion motions have little or no effect on creating or damping a local net charge. The basic point is simply that a deviation from charge neutrality is immediately converted into electron plasma oscillations and then quickly obliterated by electron–ion collisions and by Landau damping. Hence the large-scale bulk dynamics of astrophysical plasma is effectively charge neutral throughout, and there are no large-scale electrostatic effects.

Appendix C Imposition of a Large-scale Electrostatic Field

Consider what happens in a collisionless plasma when a large-scale electric field **E** is somehow impressed across the plasma. Such a field can be created in a plasma confined in a magnetic field in the laboratory using a powerful external energy source, e.g., a condensor bank, but, of course, that is not a circumstance arising in nature.

Appendix B tells us that in the absence of a magnetic field the mobile electrons rush to neutralize the electric field in the characteristic time $1/\omega_p$. In the presence of a magnetic field **B**, the component $\mathbf{E}_{\parallel}$ of the electric field parallel to **B** is neutralized in the same way, while the remaining perpendicular component meets a different fate. To show what happens suppose that an electric field $E(t)$ in the y direction is applied across the uniform magnetic field B in the z direction. The motion of an electron in those crossed fields is described by

$$\frac{d^2x}{dt^2} = -\Omega\frac{dy}{dt} \qquad \frac{d^2y}{dt^2} = -\frac{e}{m}E(t) + \Omega\frac{dx}{dt} \tag{C1}$$

where $\Omega = eB/mc$ is the electron cyclotron frequency. It follows that

$$\frac{d^3x}{dt^3} + \Omega^2\frac{dx}{dt} = \frac{e\Omega}{m}E(t) \tag{C2}$$

The general solution to this equation can be written

$$\frac{dx}{dt} = +\frac{e}{m}\int_0^t d\tau \sin\Omega(t - \tau)E(\tau) + Q_1 \sin\Omega t + Q_2 \cos\Omega t \tag{C3}$$

where Q_1 and Q_2 are arbitrary constants.

Suppose, then, that the electron is at rest and the constant electric field E is suddenly introduced at time $t = 0$. Then, for $t > 0$, the subsequent electron velocity is

$$\frac{dx}{dt} = c\frac{E}{B}(1 - \cos\Omega t) \tag{C4}$$

The electron moves in a circle with the cyclotron frequency Ω while the center of the circle (the so called *guiding center*) moves with the velocity $V = cE/B$ in the x direction. This is, of course, the familiar electric drift velocity $c\mathbf{E}\times\mathbf{B}/B^2$, which is independent of the charge and mass of the particle. The same result, then, applies to ions, the only difference being the much smaller ion cyclotron frequency eB/Mc. The ions and electrons find themselves circling the magnetic field in opposite directions at their respective cyclotron frequencies while they both have the same electric drift velocity across the magnetic field.

The essential point is that there is no electric field in the frame of reference moving with the electric drift velocity V in the x direction. This is readily shown from the Lorentz transformation (7.1) with $\mathbf{v} = c\mathbf{E}\times\mathbf{B}/B^2$. We can see that this must be true or the particle would not be moving in that reference frame, because, with a nonvanishing electric field in that frame, there would be a further electric drift velocity, putting the plasma into some other moving frame.

Note that upon switching on the field E at time $t = 0$, there is a transient rush of electric current parallel to E as the ions and electrons are initially accelerated in opposite directions. It is the Lorentz force of that motion that accelerates both electrons and ions into the electric drift velocity V and out of the electric field E. So equal numbers of electrons and ions represent a plasma with a bulk velocity equal to the local electric drift velocity. Or, given a plasma with a bulk velocity $\mathbf{v}$, and noting that there is no electric field in the frame of that plasma, it follows that there is the electric field $\mathbf{E} = -\mathbf{v}\times\mathbf{B}/c$ in the laboratory. For, even if we somehow artificially introduce an electric field perpendicular to $\mathbf{B}$, the plasma evades that electric field by accelerating into the frame in which there is no electric field.

As a final example, suppose that the electric field $E(t)$ is turned on slowly, over times long compared to the cyclotron period. Consider the case that the electron is initially at rest at time $t = -\infty$ while, say, $E(t) = E \exp \varepsilon t$, where $\varepsilon \ll \Omega$. It follows that

$$\frac{dx}{dt} = +\frac{eE}{m}\int_{-\infty}^{t} d\tau \exp \varepsilon\tau \sin \Omega(t - \tau)$$

$$= +c\frac{E \quad \exp \varepsilon t}{B(1 + \varepsilon^2/\Omega^2)} \tag{C5}$$

There is no cyclotron motion about the drifting guiding center, and, neglecting terms second order in ε/Ω compared to one, the guiding center moves in the frame of reference in which there is no electric field.

Suppose, then, that the electric field ceases to grow at time $t = 0$, remaining thereafter at the fixed value E. Going back to eqn. (C3), it follows that at time $t = 0$ we must have $Q_1 = 0$ and

$$Q_2 = \frac{cE}{B(1 + \varepsilon^2/\Omega^2)}$$

Combining this with the particular integral (C4) for $dx/dt = 0$ at $t = 0$, we have

$$\frac{dx}{dt} = c\frac{E}{B}\left(1 - \frac{\varepsilon^2/\Omega^2}{1 + \varepsilon^2/\Omega^2}\cos\Omega t\right)$$

The particle has only a slight circular motion about the moving guiding center when $\varepsilon \ll \Omega$.

In summary, the introduction of an electric field into a given frame of reference accelerates the particles into the frame of reference in which there is no electric field, and the cyclotron motion around the guiding center in that frame of reference depends on how suddenly the electric field is introduced. An abrupt switch-on provides a cyclotron motion equal to the electric drift velocity cE/B; a slow introduction of the electric field produces virtually no cyclotron motion.

Appendix D Electric Charge Density in an Electric Current

In the real world the electron mass m is small compared to the ion mass M. So the mean bulk plasma velocity $\mathbf{v}$ is essentially the same as the ion velocity, and the mean electron velocity is $\mathbf{v} + \mathbf{u}$, where $\mathbf{u}$ is the mean electron conduction velocity. Denote the ion number density by N in the frame of reference of the ions, moving with velocity $\mathbf{v}$. Then, assuming no net electric charge in the system, the number density of the electrons is also N if measured in the reference frame of the electrons, moving with velocity $\mathbf{v} + \mathbf{u}$. Therefore, in the laboratory frame the ion number density N_i and the electron number density N_e are given by

$$\begin{aligned} N_i &= \frac{N}{(1 - v^2/c^2)^{1/2}} \\ &\cong N\left(1 + \frac{v^2}{2c^2} + \cdots\right) \end{aligned} \tag{D1}$$

and

$$\begin{aligned} N_e &= \frac{N}{\left(1 - |\mathbf{v} + \mathbf{u}|^2/c^2\right)^{1/2}} \\ &\cong N\left(1 + \frac{|\mathbf{v} + \mathbf{u}|^2}{2c^2} + \cdots\right) \end{aligned} \tag{D2}$$

To keep the signs straight, suppose that $\mathbf{v} \cdot \mathbf{j} > 0$. The electron conduction velocity $\mathbf{u}$ is somewhere in the direction opposite to $\mathbf{v}$, so that $|\mathbf{v} + \mathbf{u}|^2 = (v - u)^2$. The net charge density is, then,

$$\begin{aligned} \delta &= e(N_i - N_e) \\ &\cong -\frac{eN}{c}\left(\mathbf{v} \cdot \mathbf{u} + \tfrac{1}{2}u^2\right) \end{aligned} \tag{D3}$$

in agreement with eqn. (7.5) upon recognizing that $u << v$. In fact, the additional term $u^2/2$ remains when $\mathbf{v} = 0$ because the electrons are drifting with the small conduction velocity $\mathbf{u}$ relative to the plasma.

To get an idea of just how small u/v might be, suppose that v is of the same order as the Alfvèn speed $C = B/(4\pi NM)^{1/2}$. Let l represent the characteristic scale of the large-scale variation of the magnetic field **B** so that $\nabla \times \mathbf{B}$ can be written B/l in order of magnitude. It follows from Ampere's law that

$$u = -\frac{cB}{4\pi Nel}$$

in order of magnitude. Hence,

$$\frac{u^2}{v^2} = \frac{c^2}{l^2}\left(\frac{M}{4\pi Ne^2}\right)$$

and

$$\frac{u}{v} = \frac{c}{l\omega_i}$$

in order of magnitude, where $\omega_i = (4\pi Ne^2/M)^{1/2}$ is the ion plasma frequency, and c/ω_i is the ion inertial length. Then denote the ion thermal velocity perpendicular to the magnetic field **B** by $\mathbf{w}_\perp$, so that the thermal ion cyclotron radius R is $Mw_\perp c/eB$. It follows that the ion inertial length is $c/\omega_i = R\ (C/w_\perp)$. Thus, the ion inertial length is equal to the cyclotron radius when the thermal velocity is equal to the Alfvèn speed. So, when $w_\perp$ and v are of the same general order as the Alfvèn speed, we have u/v of the order of R/l, which is small in the large-scale magnetic fields with which we are concerned. On the other hand, if **v** is very small, falling below the electron conduction velocity **u**, then eqn. (D3) provides the correct charge density, given by eqn. (7.5).

Appendix E The Transverse Invariant w_n^2/B

The motion of a charged particle of mass M in a large-scale magnetic field **B** preserves the quantity w_n^2/B, where w_n is the particle velocity perpendicular to **B**. This is readily shown in the stationary magnetic field by noting that the total particle velocity u is constant over time. Hence, if u_s is the velocity parallel to B, it follows that

$$u^2 = u_s^2 + w_n^2 \tag{E1}$$

Equation (9.53) can be rewritten as

$$\frac{d}{ds}u_s^2 = -\frac{w_n^2}{B}\frac{dB}{ds} \tag{E2}$$

Since u is constant, it follows that $du_s^2/ds = -dw_n^2/ds$, so that eqn. (E2) can be rewritten as

$$\frac{dw_n^2}{ds} = +\frac{w_n^2}{B}\frac{dB}{ds} \tag{E3}$$

requiring that w_n^2 is proportional to B, i.e., w_n^2/B does not vary as the particle moves along the field.

Suppose, then, that the field B is uniform along the field but varies slowly with time. The cyclotron radius $R = Mw_n c/eB(t)$ varies with time. The Lorentz force exerted on the particle is equal to the centrifugal force Mw_n^2/R, so the Lorentz force does the work dW on the particle in the amount

$$dW = -\frac{Mw_n^2}{R}dR \tag{E4}$$

for a cyclotron radius change dR. The work increases the kinetic energy $Mw_n^2/2$ in the amount

$$\frac{dMw_n^2/2}{Mw_n^2/2} = -2\frac{dR}{R} \tag{E5}$$

With $R = Mw_n c/eB$ it follows that

$$\frac{dR}{R} = \frac{dw_n}{w_n} - \frac{dB}{B}$$

$$= \frac{dMw_n^2/2}{2(Mw^2/2)} - \frac{dB}{B} \tag{E6}$$

Combining eqns. (E5) and (E6) eliminates dR/R and yields eqn. (E3) again, from which it follows that w_n^2/B is constant in the present case of a time dependent field. The variation of R represents that two dimensional adiabatic compression of the cyclotron motion, for which the temperature, i.e., w_n^2, varies in proportion to N.

The variation of the field B with distance s along the field and the variation with time can be combined by supposing that for an infinitesimal time Δt the magnetic field varies with t but not with s, followed by a time interval Δt in which the field varies only with s. Repeat this two-step process infinitely many times in a finite period, from which it follows that w_n^2/B is invariant in magnetic fields varying over both space and time.

It is interesting to note, then, that the angular momentum $MRw_n = (M^2c/q)(w_n^2/B)$ of the cyclotron motion around the guiding center is invariant. Then note that the flux Φ enclosed by the cyclotron orbit is

$$\Phi = \pi R^2 B$$

$$= \frac{\pi M^2 c^2}{e^2}\left(\frac{w_n^2}{B}\right) \tag{E7}$$

and obviously an invariant. That is to say, the orbit encloses a fixed amount of magnetic flux Φ as the particle moves in the inhomogeneous time-varying field.

Appendix F Blocking the Flow of Electric Current

To show formally what happens when the electric current, given by Ampere's law, eqn. (3.2), is blocked by an electrically insulating barrier, consider the simple uniformly twisted flux tube with its axis oriented along the z axis, described in section 11.3. Then if ϖ represents radial distance from the axis of the tube, the initial magnetic field is given by

$$B_\varpi = 0 \qquad B_\varphi = B_\varphi(\varpi) \qquad B_z = B_z(\varpi). \tag{F1}$$

The pressure of the incompressible fluid is such as to maintain radial equilibrium.

Now, as described in section 11.3, at time $t = 0$ the electrical conductivity is turned off throughout the region $-\varepsilon < z < +\varepsilon$ so that the electric current in this thin region falls abruptly to zero. The azimuthal magnetic field B_φ radiates away as a free electromagnetic wave, leaving kinks in the magnetic field at $z = \pm\varepsilon$, as illustrated in Fig. 11.2b. The magnetic stress at the kinks introduces an azimuthal velocity v_φ, whose time variation is described by the momentum equation

$$\rho\frac{\partial v_\varphi}{\partial t} = \frac{B_z}{4\pi}\frac{\partial B_\varphi}{\partial z} \tag{F2}$$

while the radial and longitudinal momentum equations become

$$\rho\frac{\partial v_\varpi}{\partial t} = -\frac{\partial}{\partial\varpi}\left(p + \frac{B_\varphi^2 + B_z^2}{8\pi}\right) + \frac{\rho v_\varphi^2}{\varpi} - \frac{B_\varphi^2}{4\pi\varpi} \tag{F3}$$

$$\rho\frac{\partial v_z}{\partial t} = -\frac{\partial}{\partial z}\left(p + \frac{B_\varphi^2}{8\pi}\right) \tag{F4}$$

The uniform fluid density is denoted by ρ and the fluid pressure satisfies

$$p + \frac{B_\varphi^2 + B_z^2}{8\pi} = \text{constant} \tag{F5}$$

with $\partial B_z/\partial z = 0$. For the usual torsional wave we would have

$$\rho v_\varphi^2(\varpi, t) = \frac{B_\varphi^2(\varpi, t)}{4\pi} \tag{F6}$$

The momentum equations reduce to the eqn. (F4) for $\partial v_\varphi/\partial t$ with $\partial v_\varpi/\partial t = \partial v_z/\partial t = 0$. This is the familiar torsional wave propagating along $B_z(\varpi)$ at the Alfvèn speed $C(\varpi) = B_z(\varpi)/(4\pi\rho)^{1/2}$.

The waves in which we are interested are the kinks formed at $z = \pm\varepsilon$ at time $t = 0$ and propagating away along the twisted flux bundle. Thus, the passage of the wave switches off the preexisting $B_\varphi(\varpi)$, leaving $B_\varphi = 0$ behind the advancing wave front, with the magnetic energy $B_\varphi^2/8\pi$ converted into the circular motion v_φ. It follows that $\rho v_\varphi^2(\varpi)/2$ behind the advancing wave front is equal to $B_\varphi^2(\varpi)/8\pi$ ahead of the wave front, while $B_\varphi = 0$ behind and $v_\varphi = 0$ ahead. Note, then, that the initial equilibrium condition ahead of the wave front is

$$\frac{d}{d\varpi}\left(p + \frac{B_\varphi^2 + B_z^2}{8\pi}\right) + \frac{B_\varphi^2}{4\pi\varpi} = 0$$

The magnetic stresses in B_φ set the fluid in the azimuthal motion described by eqn. (F6), replacing the inward magnetic force $B_\varphi^2/4\pi\varpi$ with the outward centrifugal force $\rho v_\varphi^2/\varpi$ and a corresponding compensating change in the fluid pressure p across the wave front. The essential point is that $B_\varphi^2/4\pi\varpi$ and $\rho v_\varphi^2/\varpi$ do not cancel each other at each point, as they do in the ordinary torsional Alfvèn wave. That is to say, the form of the wave front evolves as it propagates as a consequence of the pressure jump across the wave front. This effect is, of course, second order in the wave amplitude B_φ, and in the limit of $B_\varphi \ll B_z$ the linearized equations give a wave propagating without evolution. The wave remains the simple kink sketched in Fig. 11.2b.

The linear case is sufficient for the present discussion, and we note that the wave front at the radius ϖ advances at the Alfvèn speed

$$C(\varpi) = \pm\frac{B_z(\varpi)}{(4\pi\rho)^{1/2}}$$

jerking the fluid into the azimuthal motion v_φ as the front sweeps by. This follows directly from the MHD equations. The fluid velocity is initially zero. Then, noting that $dz/dt = C(\varpi)$, or $dt = dz/C(\varpi)$, integrate eqn. (F2) over time across the wave front, yielding the velocity

$$v_\varphi(\varpi) = \int dt\frac{\partial v_\varphi}{\partial t}$$

$$= \frac{1}{\rho}\int \frac{dz}{C(\varpi)}\frac{B_z(\varpi)}{4\pi}\frac{\partial B_\varphi}{\partial z}$$

$$= \frac{1}{(4\pi\rho)^{1/2}} \int dz \frac{\partial B_\varphi}{\partial z}$$

$$= \frac{B_\varphi(\varpi)}{(4\pi\rho)^{1/2}}$$

upon passage of the wave front.

Now the azimuthal magnetic field is initially equal to $B_\varphi(\varpi)$ and falls to zero with the passage of the wave front. It is readily shown from the induction equation for B_φ that

$$\frac{\partial B_\varphi}{\partial t} = B_z \frac{\partial v_\varphi}{\partial z}$$

so that the change in B_φ across the wave front is

$$\begin{aligned} \Delta B_\varphi(\varpi) &= \int dt \frac{\partial B_\varphi}{\partial t} \\ &= B_z(\varpi) \int \frac{dz}{C(\varpi)} \frac{\partial v_\varphi}{\partial z} \\ &= (4\pi\rho)^{1/2} v_\varphi(\varpi) \Big|_{C(\varpi)}^{0} \\ &= -B_\varphi(\varpi) \end{aligned}$$

showing that the azimuthal magnetic field declines from the initial $B_\varphi(\varpi)$ to zero with the passage of the front.

The essential point here is the simple fact that blocking the current creates torsional waves whose kinetic energy takes over the magnetic energy associated with the initial electric current. That is to say, the MHD system avoids the inductive emf that comes with the electric current analog, creating waves instead. The electric current, blocked from its flow along the flux bundle simply takes a short cut across the radius from $+I$ along the core of the bundle to the $-I$ near the periphery. The current path is not fixed, as it is in the electric circuit analog, and the fluid is set in motion by the magnetic stresses in such a way as to avoid the electric field **E** in the laboratory frame of reference.

References

Aharonov, Y. and D. Bohm 1959, Significance of electromagnetic potentials in the quantum theory, *Phys. Rev.* **115**, 485–491.
Alfvèn, H. 1966, *Worlds and Antiworlds: Antimatter in Cosmology*, W. H. Freeman, San Francisco.
Alfvèn, H. and P. Carlquist 1967, Currents in the solar atmosphere and a theory of solar flares, *Solar Phys.* **1**, 220–228.
Alfvèn, H. and C. G. Fälthammer 1963, *Cosmical Electrodynamics, Fundamental Principles*, Clarendon Press, Oxford, UK, pp. 195–198.
Bartlett, D. F. 1990, Conduction current and the magnetic field in a circular capacitor, *Am. J. Phys.* **58**, 1168–1172.
Batchelor, G. K. 1967, *An Introduction to Fluid Dynamics*, Cambridge University Press, Cambridge, UK.
Batchelor, G. K. 1971, *The Theory of Homogeneous Turbulence* (2nd ed.), Cambridge University Press, Cambridge, UK.
Biskamp, D. 1986, Magnetic reconnection via current sheets, *Phys. Fluids* **29**, 1520–1531.
Biskamp, D. and J. F. Drake 1994, Dynamics of the sawtooth collapse in Tokomak plasmas, *Phys. Rev. Lett.* **73**, 971–974.
Biskamp, D., E. Schwartrz, and J. F. Drake 1997, Two-fluid theory of collisionless plasma reconnection, *Phys. Plasmas* **4**, 1002–1009.
Biswas, T. 1988, Displacement current—a direct derivation, *Am. J. Phys.* **56**, 373–374.
Bittencourt, J. A. 1986, *Fundamentals of Plasma Physics*, Pergamon Press, New York, Chap. 12.
Born, M. and E. Wolf 1975, *Principles of Optics* (5th ed.), Pergamon Press, Oxford, UK, Chap. 3.
Braginsky, V. B. and V. I. Panov 1971, Verification of the equivalence of inertial and gravitational masses, *Zh. Eksp. Teor. Fiz* **61**, 873–879; English translation in *Sov. Phys. JETP* **34**, 464–466.
Chapman, S. 1954, The viscosity and thermal conductivity of a completely ionized gas, *Astrophys. J.* **120**, 151–155.
Chew, G. F., M. L. Goldberger, and F. E. Low 1956, The Boltzman equation and the one-fluid hydromagnetic equations in the absence of collisions, *Proc. R. Soc. London Ser. A*, **236**, 112–118.
Drake, J. F., R. G. Kleva, and M. E. Mandt 1994, Structure of thin current layers: implications for magnetic reconnection, *Phys. Rev. Lett.* **73**, 1251–1252.
Drake, J. F., M. A. Shay, W. Thongthai, and M. Swisdak 2005, Production of energetic electrons during magnetic reconnection, *Phys. Rev. Lett.* **94**, 095001, 1–4.
Dylla, H. F. and J. G. King 1973, Neutrality of molecules by a new method, *Phys. Rev. A* **7**, 1224–1229.

Ehrenberg, W. and R. E. Siday 1949, The refractive index in electron optics and the principles of dynamics, *Proc. Phys. Soc. (London)* **B62**, 8–21.

Eötvös, R. V., D. Pekár, and E. Fekete 1922, Beiträge zum Gesetze der Proportionalität von Trägheit und Gravität, *Ann. Phys.* (Germany) **18**, 11–66.

Feldman, U., J. M. Laming, P. Mandelbaum, W. H. Goldstein, and A. Osterheld 1992, A burst model for line emission in the solar atmosphere, II: coronal extreme ultraviolet, *Astrophys. J.* **398**, 692–297.

French, A. P. and J. R. Tessman 1963, Displacement currents and magnetic fields, *Am. J. Phys.* **31**, 201–204.

Gaizauskas, V., K. L. Harvey, J. W. Harvey, and C. Zwaan 1983, Large-scale patterns formed by solar active regions during the ascending phase of cycle 21, *Astrophys. J.* **265**, 1056–1065.

Goldstein, S. 1938, *Modern Developments in Fluid Dynamics*, Clarendon Press, Oxford, UK (reprinted by Dover Publications, New York, 1965).

Hoyle, F. 1960, On the possible consequences of a variability of the elemental charge, *Proc. R. Soc. London Ser. A* **257**, 431–442.

Jemella, B. D., M. A. Shay, J. F. Drake, and B. N. Rogers 2003, Impact of frustrated singularities on magnetic island reconnection, *Phys. Rev. Lett.* **91**, 125002, 1–4.

Ji, H., T. Carter, S. Hsu, and M. Yamada 2001, Study of local reconnection physics in a laboratory plasma, *Earth, Planets and Space* **53**, 539–545.

Katsukawa, Y. 2003, Spatial and temporal extent of solar nanoflares and their energy range, *Pub. Astron. Soc. Jpn.* **55**, 1025–1031.

Katsukawa, Y. and S. Tsuneta 2005, Magnetic properties at footpoints of hot and cool loops, *Astrophys. J.* **621**, 498–511.

Klimchuk, J. A. 2006, On solving the coronal heating problem, *Solar Phys.* (in press).

Kolmogoroff, A. N. 1941a, The local structure of turbulence in incompressible viscous fluid for very large Reynolds numbers, *Dokl. Akad. Nauk. SSSR* **30**, 301–305.

Kolmogoroff, A. N. 1941b, Dissipation of energy in locally isotropic turbulence, *Dokl. Akad Nauk. SSSR* **32**, 82–85.

Kolmogoroff, A. N . 1962, A refinement of previous hypotheses concerning the local structure of turbulence in a viscous incompressible fluid at high Reynolds number, *J. Fluid Mech.* 13, 82–85.

Kopecky, M. and G. V. Kuklin 1969, On a more precise calculation of the electric conductivity in the photosphere and in sunspots, *Solar Phys.* **6**, 241–250.

Kopecky, M. and V. Obridko 1968, On the energy release by magnetic field dissipation in the solar atmosphere, *Solar Phys.* **5**, 354–358.

Kraichnan, R. H. 1974, On Kolmogoroff's inertial range theorie *J. Fluid Mech.* **62**, 305–330.

Kraichnan, R. H. and D. Montgomery 1980, Two dimensional turbulence, *Rep. Prog. Phys.* **43**, 549–619.

Lamb, H. 1932, *Hydrodynamics* (6th ed.), Dover, New York.

Landau, L. D. and E. M. Lifschitz 1959, *Fluid Mechanics*, Addison-Wesley, Reading, MA.

Low, B. C. 1990, Equilibrium and dynamics in coronal magnetic fields, *Annu. Rev. Astron. Astrophys.* **28**, 491–524.

Low, B. C. and M. A. Berger 2002, A morphological study of helical coronal magnetic structures, *Astrophys. J.* **689**, 644–657.

Lyttleton, R. A. and H. Bondi 1959, On the physical consequences of a general excess charge, *Proc. R. Soc. London Ser. A* **252**, 313–333.

Lyttleton, R. A. and H. Bondi 1960, Note on the preceding paper [Hoyle 1960], *Proc. R. Soc. London Ser. A* **257**, 442–444.

Mandt, M. E., R. E. Denton, and J. F. Drake 1994, Transition to whistler mediated magnetic reconnection, *Geophys. Res. Lett.* **21**, 73–76.

McComb, W. D. 1990, *The Physics of Fluid Turbulence*, Clarendon Press, Oxford, UK.

Mello, P. A. 1972, A remark on Maxwell's displacement current, *Am. J. Phys.* **40**, 1010–1013.

Melrose, D. B. 1995, Current paths in the corona and energy release in solar flares, *Astrophys. J.* **451**, 391–401.

Mihalas, D. and B. Weibel-Mihalas 1984, *Foundations of Radiation Hydrodynamics*, Oxford University Press, Oxford, UK.

Misner, C. W., K. S. Thorne, and J. A. Wheeler 1973, *Gravitation*, W. H. Freeman, San Francisco, p.13.

Olariu, S. and I. I. Popescu 1985, The quantum effects of electromagnetic fluxes, *Rev. Mod. Phys.* **57**, 339–436.

Parker, E. N. 1957a, Newtonian development of the dynamical properties of ionized gases at low densities, *Phys. Rev.* **107**, 924–933.

Parker, E. N. 1957b, Sweet's mechanism for merging magnetic fields in conducting fluids, *J. Geophys. Res.* **62**, 509–520.

Parker, E. N. 1958, Dynamics of the interplanetary gas and magnetic fields, *Astrophys. J.* **128**, 664–676.

Parker, E. N. 1963a, *Interplanetary Dynamical Processes*, Wiley-Interscience, New York, p. 151.

Parker, E. N. 1963b, Solar-flare phenomena and the theory of reconnection and annihilation of magnetic fields, *Astrophys. J. Suppl.* **8**, 177–212.

Parker, E. N. 1967, Confinement of a magnetic field by a beam of iuons, *J. Geophys. Res.* **72**, 177–212.

Parker, E. N. 1970, The origin of magnetic fields, *Astrophys. J.* **160**, 383–404.

Parker, E. N. 1971a, The generation of magnetic fields in astrophysical bodies, II: the galactic magnetic field, *Astrophys. J.* **163**, 255–278.

Parker, E. N. 1971b, The generation of magnetic fields in astrophysical bodies, IV: the solar and terrestrial dynamos, *Astrophys. J.* **164**, 491–509.

Parker, E. N. 1971c, The generation of magnetic fields in astrophysical bodies, VI: periodic modes of the galactic magnetic fields, *Astrophys. J.* **166**, 295–300.

Parker, E. N. 1972, Topological dissipation and the small-scale fields in turbulent gases, *Astrophys. J.* **174**, 499–510.

Parker, E. N. 1979, *Cosmical Magnetic Fields*, Clarendon Press, Oxford, UK, Chap. 11.

Parker, E. N. 1981a, The dissipation of inhomogeneous magnetic fields and the problem of coronae, I: dislocation and flattening of flux tubes, *Asytrophys. J.* **244**, 631–643.

Parker, E. N. 1981b, The dissipation of inhomogeneous magnetic fields and the problem of coronae, II: the dynamics of dislocated flux tubes, *Astrophys. J.* **244**, 644–652.

Parker, E. N. 1983, Magnetic neutral sheets in evolving fields, II: formation of the solar corona, *Astrophys. J.* **264**, 642–647.

Parker, E. N. 1984, Galactic magnetic fields and magnetic monopoles, in *Monopole '83*, Plenum, New York, pp. 125–136.

Parker, E. N. 1986, Equilibrium of magnetic fields with arbitrary interweaving of the lines of force, I: discontinuities in the torsion, *Geophys. Astrophys. Fluid Dyn.* **34**, 243–264.

Parker, E. N. 1987, Magnetic monopole plasma oscillations and the survival of galactic magnetic fields, *Astrophys. J.* **321**, 349–354.

Parker, E. N. 1988, Nanoflares and the solar X-ray corona, *Astrophys. J.* **330**, 474–479.

Parker, E. N. 1989a, Spontaneous tangential discontinuities and the optical analogy for static magnetic fields, I: the optical analogy, *Geophys. Astrophys. Fluid Dyn.* **45**, 169–182.

Parker, E. N. 1989b, Spontaneous tangential discontinuities and the optical analogy for static magnetic fields, III: zones of exclusion, *Geophys. Astrophys. Fluid Dyn.* **46**, 105–133.

Parker, E. N. 1991, The optical analogy for vector fields, *Phys. Fluids B*, **3**, 2652–2659.

Parker, E. N. 1992, Fast dynamos, cosmic rays, and the galactic magnetic field, *Astrophys. J.* **401**, 137–145.

Parker, E. N. 1994, *Spontaneous Current Sheets in Magnetic Fields*, Oxford University Press, New York.

Parker, E. N. 1996a, The alternative paradigm for magnetospheric physics, *J. Geophys. Res.* **101**, 10587–10625.

Parker, E. N. 1996b, Comment on "Current paths in the corona and energy release in solar flares," *Astrophys. J.* **471**, 489–496.

Parker, E. N. 2001, Solar activity and classical physics, *Chinese J. Astron. Astrophys.* **1**, 99–124.

Parks, G. K. 2004, Why space physics needs to go beyond the MHD box, *Space Sci. Rev.* **113**, 97–125.

Petschek, H. E. 1964, Magnetic field annihilation, in *The Physics of Solar Flares*, Scientific and Technical information Division, National Aeronautics and Space Administration, Washington DC, NASA Sp-50, pp. 425–439.

Petschek, H. E. and R. M. Thorne, 1967, The existence of intermediate waves in neutral sheets, *Astrophys. J.* **147**, 1157–1163.

Red Rock Country Visitors Guide, Fall 2004, A publication of the *Sedona Red Rock News*, p. 8.

Reynolds, O. 1883, An experimental investigation of the circumstances which determine whether the motion of water shall be direct or sinuous, and the law of resistance in parallel channels, *Philos. Trans. R. Soc. London Ser. A* **174**, 935–982.

Reynolds, O. 1895, On the dynamical theory of incompressible viscous fluids and the determination of the criterion, *Philos. Trans. R. Soc. London Ser. A* **186**, 123–164.

Rogers, B. N., R. E. Denton, J. F. Drake, and M. A. Shay 2002, Role of dispersive waves in collisionless magnetic reconnection, *Phys. Rev. Lett.* **87**, 195004, 1–4.

Rohrlich, D. and R. G. Chambers 1960, Topological effects of quantum mechanics, *Phys. Rev. Lett.* **5**, 3–5.

Roll, P. G., G. R. Krofkov, and R. H. Dicke 1964, The equivalence of inertial and passive gravitational mass, *Ann. Phys.* (USA) **26**, 442–517.

Saffman, P. G. 1992, *Vortex Mechanics*, Cambridge University Press, Cambridge, UK.

Samson, K. and J. Aukshunas 2002, in *Frommer's Arizona 2002*, Hungry Minds, New York, p. 167.

Schindler, K., M. Hesse, and J. Birn 1991, Magnetic field-aligned electric potentials in nonideal plasma flows, *Astrophys. J.* **380**, 293–301.

Shadowitz, A. 1988, *The Electromagnetic Field*, Dover, New York.

Shay, M. and J. F. Drake 2001, The role of electron dissipation on the rate of collisionless magnetic reconnection, *Geophys. Res. Lett.* **25**, 3759–3762.

Spitzer, L. 1956, *The Physics of Fully Ionized Gases*, Wiley-Interscience, New York.

Stratton, J. A. 1941, *Electromagnetic Theory*, McGraw-Hill, New York, p. 3.

Sturrock, P. A., W. W. Dixon, J. A. Klimchuk, and S. Antiochos 1990, Episodic coronal heating. *Astrophys. J. Lett.* **356**, L31–L34.

Sweet, P. A. 1958, The production of high energy particles in solar flares, *Nuovo Cimeno Suppl.* **8**, *Series X*, 188–196.

Tonomura, A., N. Osakabe, T. Matsuda, T. Kawasaki, J. Endo, S. Yano, and H. Yamada 1986, Evidence for Aharonov-Bohm effect with magnetic field completely shielded from electron wave, *Phys. Rev. Lett.* **56**, 792–795.

Turner, M., E. N. Parker, and T. Bogdan 1982, Magnetic monopoles and the survival of magnetic fields, *Phys. Rev. D* **26**, 1296–1305.

Van Ballegooijen, A. A. 1985, Electric currents in the corona and the existence of magnetic equilibrium, *Astrophys. J.* **298**, 421–430.

Vasyliunas, V. M. 1999, A note on current closure, *J. Geophys. Res.* **104**, 25143–25144.

Vasyliunas, V. M. 2001, Electric field and plasma flow: what drives what? *Geophys. Res. Lett.* **28**, 2177–2180.

Vasyliunas, V. M. 2005a, Time evolution of electric fields and currents and generalized Ohm's law, *Ann. Geophys.* **23**, 1347–1354.

Vasyliunas, V. M. 2005b, Relation between magnetic fields and electric currents in plasmas, *Ann. Geophys.* **23**, 2589–2597.

Vasyliunas, V. M. and P. Song 2005, Meaning of Joule heating, *J. Geophys. Res.* **110**, (A02301)doi:10.1029/2004JA010615.

Wang, X. and A. Bhattacharjee 1992, Forced reconnection and current sheet formation in Taylor's model, *Phys. Fluids B* **4**, 1795–1799.
Wilczek, F. 2005, On absolute units, I: choices, *Physics Today*, October, 12–13.
Wilczek, F. 2006, On absolute units, II: challenges and responses, *Physics Today*, January, 10–11.
Wu, T. Y. 1966, *Kinetic Equations of Gases and Fields*, Addison-Wesley, Reading, MA.
Wu, T. T. and C. N. Yang 1975, Concept of nonintegrable phase factors and global formulation of gauge fields, *Phys. Rev. D* **12**, 3845–3857.
Yamada, M. 2001, Review of the recent controlled experiments for study of local reconnection physics, *Earth, Planets, and Space* **53**, 509–519.
Yamada, M. et al. 2000, Experimental investigation of the neutral sheet profile during magnetic reconnection, *Phys. Plasmas* **7**, 1781–1787.
Yamada, M., Y. Ren, H. Ji, J. Breslau, S. Gerhardt, R. Kulsrud, and A. Kuritsyn 2006, Experimental study of two-fluid effects on magnetic reconnection in a laboratory plasma with variable collisionality, *Phys. Plasmas* (in press).

Index

Milton Keynes UK
Ingram Content Group UK Ltd.
UKHW051131310824
447656UK00017B/233

9 780691 128412